PCT, Spin and Statistics, and All That

PCT, Spin and Statistics, and All That

R.F. STREATER
Kings College London

A.S. WIGHTMAN
Princeton University

Princeton University Press
Princeton and Oxford

Published by Princeton University Press, 41 William Street,
Princeton, New Jersey 08540
In the United Kingdom: Princeton University Press,
3 Market Place, Woodstock, Oxfordshire OX20 1SY

First paperback printing, with revised Preface and corrections, 2000

The publisher is pleased to acknowledge the assistance of Cecelia Duray-Bito,
who produced the illustrations.

Originally published in 1964 as part of the Mathematical Physics Monograph series,
by W.A. Benjamin, Inc.

PCT, Spin and Statistics, and All That

First printing, 1964

Second printing, with additions and corrections, 1978

Third printing, 1980

Library of Congress Cataloging-in-Publication Data

Streater, R. F.
PCT, spin and statistics, and all that / R.F. Streater, A.S. Wightman.
p. cm.
"Originally published in 1964 as part of the Mathematical physics monograph series, by W.A. Benjamin"—T.p.
Includes bibliographical references and index.
ISBN 0-691-07062-8
1. Quantum field theory. I. Wightman, A. S. II. Title.

QC174.45 .S87 2000
530.14'3—dc21 00-061116

Printed in the United States of America.

The paper used in this publication meets the minimum
requirements of ANSI/NISO Z39.48-1992 (R1997) (*Permanence of Paper*)

ABCDEFGHIJ-AL-89

www.pup.princeton.edu

10 9 8 7 6 5 4 3 2

ISBN-13: 978-0-691-07062-9 (pbk.)

ISBN-10: 0-691-07062-8 (pbk.)

Contents

Preface

The idea of this book arose in a conversation with H.A. Bethe, who remarked that a little book about modern field theory which contained only Memorable Results would be a Good Thing. In the field of historical research this approach led to the publication of a treatise† which has become a standard text for serious students. Although it is often dangerous to use the tried and true methods of one subject in another field of research, the application to physics of the principles of that book has led to at least one good result: we have eliminated all theorems whose proofs are non-existent.

R.F. Streater
A.S. Wightman
November 1963

In the 1978 edition of this book, we added Appendix A, which outlined three significant developments in the general theory of quantized fields that had occurred since the appearance of the first edition: constructive quantum field theory, the theory of local algebras, and the theory of superselection rules.

Neither of these first two editions contained an account of the Haag-Ruelle collision theory. For that the reader of the first edition was referred to the then forthcoming excellent book by R. Jost (*The General Theory of Quantized Fields,* American Mathematical Society, 1965). For the 1978 edition this reference was supplemented by a reference to the treatise *Introduction to Axiomatic Quantum Field Theory* by N. Bogolubov, A. Logunov and I. Todorov, W.A. Benjamin (1975).

For the present Princeton University Press reissue (2000), we want to add a

†W.C. Sellar and R.J. Yeatman, *1066 and All That*, Dutton, New York, 1931.

third recommendation to this list: H. Araki's *Mathematical Theory of Quantum Fields,* Oxford University Press, 2000. In other respects the Princeton University Press edition differs from the 1978 edition mainly in the correction of a few misprints.

R.F. Streater
A.S. Wightman

PCT, Spin and Statistics, and All That

INTRODUCTION

In the beginning, when Dirac, Jordan, Heisenberg, and Pauli created the quantum theory of fields, it was not expected that it would provide a consistent description of Nature. After all, it was only a quantized version of the classical theory of Maxwell and Lorentz, a theory which was well known to be afflicted with diseases arising from the infinite electromagnetic inertia of point particles. Many physicists were of the opinion that any project to make the theory's mathematical foundation more rigorous was probably ill-advised; first the classical foundation should be set right. Such alterations might so change the basis of the theory that a mathematically rigorous discussion of any preceding version would be entirely irrelevant. More recently, it has been suggested that the trouble is that the theory is too modest; it is not designed to predict the masses of the elementary particles or the values of the coupling constants, and should be fundamentally changed with this in view.

However, attempts to go beyond the theory foundered again and again. What successes were achieved were either phenomenological, or were due to systematic developments of the original formalism. But the quantum theory of fields never reached a stage where one could say with confidence that it was free from internal contradictions—nor the converse. In fact, the Main Problem of quantum field theory turned out to be to kill it or cure it: either to show that the idealizations involved in the fundamental notions of the theory (relativistic invariance, quantum mechanics, local fields, etc.) are incompatible in some physical sense, or to recast the theory in such a form that it provides a practical language for the description of elementary particle dynamics.

The last ten years have seen a number of attempts to meet the situation head on. (The physicists who have engaged in this kind of work are sometimes dubbed the Feldverein. Cynical observers have compared them to the Shakers, a religious sect of New England who built solid barns and led celibate lives, a non-scientific equivalent of proving rigorous theorems and calculating no cross sections.) These efforts have not yet led to a solution of the Main Problem, but they have yielded a number of by-products, very general insights into the structure of a field theory. The present book is devoted to an exposition of some of these general results, the physical ideas they embody, and the mathematics necessary for their proofs.

We have included only results which have a certain definitive character. In particular, this has resulted in the omission of a description of attempts to establish a connection with the important work of Lehmann, Symanzik, and Zimmermann and others on time-ordered and retarded functions and their connection with collision theory. A great deal more work will be necessary before these subjects can be properly understood and put on a rigorous basis. Although the connection with Lehmann, Symanzik, and Zimmermann is not yet firmly established, a rigorous collision theory (based on the axioms of Chapter 3 of the present book) has been set up by D. Ruelle along lines laid down by R. Haag. The omission of this theory from this book will be mitigated by the availability of the excellent book by R. Jost, which will contain a full account (*The General Theory of Quantized Fields*, American Mathematical Society, 1965).

The first chapter contains a summary of the transformation properties of physical states in relativistic quantum mechanics. It is assumed that the reader has had an introduction to Hilbert space and its application to the description of states in quantum mechanics. It is probably worthwhile to point out to younger readers that the simplicity of the transformation laws of the physical vacuum and one-particle states under Lorentz transformations, well known today, was buried in the quantum field theory of fifteen years ago under a kitchen midden of difficult and ambiguous formalism. The task of the first chapter is to provide a language in which physical states with simple transformation properties have a simple description. For example, the concepts of bare mass and bare vacuum need not be and are not introduced.

The second chapter is an exposition of the mathematical tools used in the following. Technical details of some proofs have been omitted but an attempt has been made to get across the main mathematical ideas. The theorems are stated precisely. The presentation presupposes nothing that an undergraduate physics major has not met.

The third chapter defines the notion of field as used in this book. It is shown that a field theory is defined by the vacuum expectation values of products of field operators. While this chapter is essentially self-contained, a brush with elementary quantum field theory at the level of, say, Part II of Schweber's book† might be a help.

In Chapter 4 the three preparatory chapters are applied to get some general theorems of quantum field theory, of which the *PCT* theorem and the theorem on the connection of spin with statistics are the best known.

† S. Schweber, *An Introduction to Relativistic Quantum Field Theory*, Harper and Row, New York, 1961.

The reader who wants to get on with the systematic discussion of quantum field theory could well begin with Chapter 3 and only go back to Chapters 1 and 2 when he finds it necessary to fill in details.

Each chapter has been equipped with a bibliography to guide the reader to relevant literature. No attempt at completeness has been made. The notation is standard: Theorem 3–1 refers to the first theorem of the third chapter, and similarly for equations. Halmos notation ▮ has been used to signify the end of a proof.

CHAPTER 1

RELATIVISTIC TRANSFORMATION LAWS

You point out that care is needed in the analysis of the representations of the Lorentz group; I promise you that I will be careful.

E. Wigner

Throughout this book, states will be described in the Heisenberg picture of quantum mechanics. The Schrödinger picture is much less convenient for the description of a relativistic theory because it treats the time coordinate on a very different footing from the space coordinates; as will be proved in Chapter 4, the other commonly used picture, the interaction picture, in general does not exist. In the Heisenberg picture, to each state of the system under consideration there corresponds a unit vector, say Φ, in a Hilbert space, $\mathscr{H}$. The vector does not change with time, whereas the observables, represented by hermitian linear operators acting on $\mathscr{H}$, in general do. The scalar product of two vectors Φ and Ψ in $\mathscr{H}$ is denoted by (Φ, Ψ), called the *transition amplitude* of the corresponding states.

Two vectors that differ only by multiplication by a complex number of modulus one describe the same state, because the results of all experiments on a state described by Ψ may be expressed in terms of the quantities

$$|(\Phi, \Psi)|^2$$

which gives the probability of finding Φ if Ψ is what you have. The set $\mathbf{\Phi}$ of vectors $e^{i\alpha}\Phi$, where α varies over all real numbers, and the norm of Φ (written $\|\Phi\|$ and defined as $[(\Phi, \Phi)]^{1/2}$) is unity, is called a *unit ray*. For brevity, we shall speak of the state $\mathbf{\Phi}$. The condition $\|\Phi\| = 1$ is obviously equivalent to the convention of normalizing the probability to unity. The preceding remarks can be summarized: *states† of a physical system are represented by unit rays.*

† By "state" we shall always mean pure state. A "mixed state" can always be formed from several states by a *classical* superposition, each entering with a certain known probability which describes our ignorance of the system.

1-1. SUPER-SELECTION RULES

Suppose that the rays which describe the states of a physical system lie in a Hilbert space $\mathscr{H}$. Does every unit ray in $\mathscr{H}$ describe a possible state of the system? The answer is, in general, no. For example, no one has ever succeeded in producing a state which is a superposition of states with different charges, Q, and it is believed that they do not occur in Nature. It also seems that every *physically realizable* state must be an eigenstate of B, the baryon number, and $(-1)^F$, where F is an even integer for states of integer spin and an odd integer for states of half odd integer spin.

The operators Q,B and $(-1)^F$ are conserved in time, but these conservation laws should be distinguished from ordinary conservation laws such as, say, that for the x-component of the angular momentum, J_x. Physically realizable states do exist which are not eigenstates of J_x, for example, states with a definite value of the z-component of the angular momentum J_z.

The operator $(-1)^F$ arises because of the invariance of the results of experiment under rotation through an angle of 2π around any axis. If ψ_1 is a state of half odd integer spin, and ψ_2 one of integer spin, then a rotation through an angle 2π takes $\alpha\psi_1 + \beta\psi_2$ into $-\alpha\psi_1 + \beta\psi_2$. These two, which are physically indistinguishable, must belong to the same ray, which is possible only if $\alpha = 0$ or $\beta = 0$.

Any statement that singles out certain unit rays as not physically realizable is called a *super-selection rule*. If there are super-selection rules in a theory, then not all hermitian operators are observables, and the superposition principle does not hold in $\mathscr{H}$. However, if Q,B and $(-1)^F$ define the only super-selection rules, we can form a linear combination of any two states with the same values of Q,B and $(-1)^F$ and get a physical state. The superposition principle then holds unrestrictedly in any subspace of $\mathscr{H}$ consisting of states belonging to given eigenvalues of Q,B and $(-1)^F$.

The super-selection rules associated with Q,B and $(-1)^F$ are known as the *charge*, *baryon*, and *univalence* super-selection rules, respectively.

To study super-selection rules of a general theory systematically, one considers the set θ of all observables of the system under consideration. Each observable determines a hermitian operator in $\mathscr{H}$, not necessarily bounded (an operator A is bounded if $\|A\Phi\| \leqslant C\|\Phi\|$ for some constant C and all $\Phi \in \mathscr{H}$). In this general case a ray is said to be *physically realizable* if the projection operator onto it is an observable.† Consider the set of all bounded operators which commute with all the

† The projection operator, E_Φ, onto a vector Φ is given by the formula

$$E_\Phi \Psi = (\Phi, \Psi)[\|\Phi\|^2]^{-1}\Phi.$$

observables; this is a set θ', called the *commutant* of θ. The limitation to bounded operators in the definition of θ' is purely a matter of convenience. In fact the operators Q and B, being unbounded, do not lie in θ', but the associated projection operators, which project onto the states of various possible values of Q and B, do.

The set θ' partly characterizes the super-selection rules present in the theory. For example, if every hermitian operator is observable, every state is physically realizable since any projection operator is hermitian. Therefore, in this case there are no super-selection rules; the set θ' consists only of multiples of the identity operator.

If we now make the hypothesis that all operators of θ' commute with each other (that is sometimes called *the hypothesis of commutative super-selection rules*), the structure of the set of physically realizable states simplifies considerably. The super-selection rules in θ' can be diagonalized simultaneously, and $\mathscr{H}$ splits up into orthogonal subspaces in which each of the operators defining a super-selection rule takes a definite value. These are called *coherent subspaces.* The observables map coherent subspaces into themselves, and the only operators which are defined on a single coherent subspace, map it into itself, and commute with all observables are constant multiples of the identity; i.e., the observables, when restricted to a single coherent subspace, form an irreducible set of operators.

There is one important case in which one can prove that the hypothesis of commutative super-selection rules holds: when there exists a complete commuting set of observables.† Any operator that commutes with all operators of such a set is a function of the operators of the set. In particular, any operator that is in θ' is a function of the observables of the set. Therefore, in this case all operators in θ' commute.

Although the observables of a particular coherent subspace are irreducible, this by no means implies that they include among them every hermitian operator. For example, in a particular coherent subspace there are normalizable states with infinite energy, and states of this kind ought not to be classified as physically realizable. Thus, the projection operator onto such a state, although hermitian, is not an observable. Nevertheless, in the following we shall assume that θ' is commutative and that every ray of a coherent subspace is physically realizable. This hypothesis is made purely for mathematical convenience. Actually, the analysis could be carried out under much more general assumptions at a cost of more effort.

† Complete commuting set is the standard Dirac terminology; it is also called a *maximal Abelian set.*

It should be emphasized that the above super-selection rules for Q,B, and $(-1)^F$, like all laws of physics, depend upon experiment. It is quite unclear at the moment whether further super-selection rules exist. For example, it may be that there are laws of lepton conservation which define super-selection rules.

1–2. SYMMETRY OPERATIONS

A *symmetry operation* (sometimes called an *invariance principle*, or simply a *symmetry*) of a physical system is a correspondence which yields for each physically realizable state $\mathbf{\Phi}$, another, $\mathbf{\Phi}'$, such that all transition probabilities are preserved:

$$|(\Phi', \Psi')|^2 = |(\Phi, \Psi)|^2. \tag{1–1}$$

It is assumed that the mapping $\mathbf{\Phi} \to \mathbf{\Phi}'$ is one to one. This means that as $\mathbf{\Phi}$ runs over all physically realizable states, so does $\mathbf{\Phi}'$, and if $\mathbf{\Phi}$ and $\mathbf{\Psi}$ are distinct, so are $\mathbf{\Phi}'$ and $\mathbf{\Psi}'$. An example of a symmetry is the operator translating the system by the four-vector a. This is represented by an operator $V(a)$ which is unitary [i.e., $(V\Phi, V\Psi) = (\Phi, \Psi)$]. Another example is the CPT operator Θ, which is anti-unitary [i.e., $(\Theta\Phi, \Theta\Psi) = \overline{(\Phi, \Psi)}$]. Incidentally, the operator Θ interchanges the coherent sub-spaces with opposite charge and baryon number. Clearly, both unitary and anti-unitary operators satisfy (1–1). In fact, all mappings $\mathbf{\Phi} \to \mathbf{\Phi}'$ lead essentially to a unique transformation $\Phi \to \Phi'$ satisfying (1–1), and this transformation is either unitary or anti-unitary (Ref. 1).

Theorem 1–1

Let $\mathbf{\Phi} \to \mathbf{\Phi}'$ be a symmetry of a physical theory satisfying the hypothesis of commutative super-selection rules.

If the symmetry leaves coherent subspaces invariant, then there exists in each coherent subspace a unitary or anti-unitary operator V such that for all physically realizable states of that subspace

$$\mathbf{\Phi}' = V\mathbf{\Phi}. \tag{1–2}$$

The operator V is uniquely determined up to a phase.

If the symmetry does not leave coherent subspaces invariant, then restricted to a coherent subspace it is a one-to-one mapping onto another coherent subspace, unitary or anti-unitary and unique up to a phase.

We shall not prove the theorem here; its essence is the following. The one-to-one ray correspondence $\boldsymbol{\Phi} \rightarrow \boldsymbol{\Phi}'$ can be induced by one of many different vector correspondences, $\Phi \rightarrow \Phi'$ in the underlying Hilbert space, but in general such a correspondence is neither linear nor anti-linear; i.e., neither

$$\alpha\Phi' + \beta\Psi' = (\alpha\Phi + \beta\Psi)'$$

nor

$$\bar{\alpha}\Phi' + \bar{\beta}\Psi' = (\alpha\Phi + \beta\Psi)'$$

holds. What the theorem says is that when the vectors of a single coherent subspace are considered there is a linear or anti-linear transformation unique up to a phase (one or the other, not both!) which yields the given correspondence between rays.

A final remark about symmetry operations. Our definition clearly makes every unitary and anti-unitary operator a symmetry operator and Theorem 1–1 shows that these are essentially the only such. How then can one system be more symmetrical than another? The answer lies in the physical interpretation of the operations. Consider for example a theory of two spinless particles with coordinate and momentum operators $\mathbf{q}_1(t)$, $\mathbf{q}_2(t)$, $\mathbf{p}_1(t)$, $\mathbf{p}_2(t)$. Then the mapping $\mathbf{q}_i(t) \rightarrow -\mathbf{q}_i(t)$, $\mathbf{p}_i(t) \rightarrow -\mathbf{p}_i(t)$ of the observables onto themselves can always be defined, but only when the system possesses symmetry under space inversion will the mapping be induced by a unitary operator V according to

$$V\mathbf{q}_j(t)V^{-1} = -\mathbf{q}_j(t), \qquad V\mathbf{p}_j(t)V^{-1} = -\mathbf{p}_j(t), \qquad j = 1,2 \quad (1\text{–}3)$$

with V independent of t.

On the other hand, it may be possible to define a unitary operator, V, which has the effect of a parity operator for the center of mass motion and total momentum, even if the theory is *not* invariant under space-inversion, i.e., even if there is no V satisfying (1–3). For example, V might satisfy

$$V\mathbf{q}_1(t)V^{-1} = -\mathbf{q}_2(t), \qquad V\mathbf{q}_2(t)V^{-1} = -\mathbf{q}_1(t)$$
$$V\mathbf{p}_1(t)V^{-1} = -\mathbf{p}_2(t), \qquad V\mathbf{p}_2(t)V^{-1} = -\mathbf{p}_1(t).$$

Then

$$V(\mathbf{q}_1(t) + \mathbf{q}_2(t))V^{-1} = -(\mathbf{q}_1(t) + \mathbf{q}_2(t))$$
$$V(\mathbf{p}_1(t) + \mathbf{p}_2(t))V^{-1} = -(\mathbf{p}_1(t) + \mathbf{p}_2(t)).$$

This example makes clear that the statement that a physical system possesses space-inversion symmetry only makes sense when one has specified what space inversion is supposed to do to the observables of the system. In the remaining sections of this chapter it will be assumed that some such specification has been made for relativity transformations. In Chapter 3 the specification will be made explicit in terms of fields.

1–3. THE LORENTZ AND POINCARÉ GROUPS

Among the most important symmetries of relativistic quantum theory are those which arise from the Lorentz transformations themselves. The next few paragraphs are devoted to establishing our notation and summarizing the main facts about these symmetries.

The Lorentz-invariant scalar product of two four-vectors $x = (x^0, x^1, x^2, x^3)$ and $y = (y^0, y^1, y^2, y^3)$ will be written

$$x \cdot y = x^0 y^0 - \mathbf{x} \cdot \mathbf{y} \equiv x^\mu g_{\mu\nu} y^\nu \equiv x^\mu y_\mu \tag{1–4}$$

with the usual summation convention on repeated indices.

Here

$$x_\mu = g_{\mu\nu} x^\nu$$

and $g^{\mu\nu} = g_{\mu\nu}$ is the μ,ν component of the matrix G,

$$G = \begin{Bmatrix} 1 & 0 & 0 & 0 \\ 0 & -1 & 0 & 0 \\ 0 & 0 & -1 & 0 \\ 0 & 0 & 0 & -1 \end{Bmatrix}.$$

A Lorentz transformation Λ is a linear transformation mapping space-time onto space-time which preserves the scalar product (1–4): $(\Lambda x)\cdot(\Lambda y) = x \cdot y$. If $(\Lambda x)^\mu = \Lambda^\mu{}_\nu x^\nu$ the (real) matrix, $\Lambda^\mu{}_\nu$, of the transformation must satisfy

$$\Lambda^\kappa{}_\mu \Lambda_{\kappa\nu} = g_{\mu\nu} \qquad \text{or} \qquad \Lambda^T G \Lambda = G, \tag{1–5}$$

where the transpose Λ^T of Λ is defined by $(\Lambda^T)^\mu{}_\nu = \Lambda^\nu{}_\mu$ and indices on Λ are lowered according to

$$\begin{aligned} \Lambda_{\kappa\nu} &= g_{\kappa\sigma} \Lambda^\sigma{}_\nu \\ &= (G\Lambda)_{\kappa\nu}. \end{aligned}$$

If Λ and M satisfy (1–5), so do ΛM and Λ^{-1}. Here

$$(\Lambda M)^\mu{}_\nu = \Lambda^\mu{}_\kappa M^\kappa{}_\nu$$

$$(\Lambda^{-1})^\mu{}_\kappa \Lambda^\kappa{}_\nu = g^\mu{}_\nu = \begin{cases} 0 & \mu \neq \nu \\ 1 & \mu = \nu, \end{cases} \tag{1–6}$$

so the Lorentz transformations form a group, the Lorentz group, L. Two Lorentz transformations Λ and M are defined to be close to one another if the numbers $\Lambda^\mu{}_\nu$ and $M^\mu{}_\nu$ are close for all $\mu,\nu = 0,1,2,3$. Clearly, with this definition, Λ^{-1} and ΛM are continuous functions of Λ and M, respectively. Furthermore, it makes sense to say that two Lorentz transformations can be connected to one another by a continuous curve of Lorentz transformations.

L has four components, each of which is connected in the sense that any one point can be connected to any other, but no Lorentz transformation in one component can be connected to another in another component. To see this, note that $\det \Lambda$ and $\operatorname{sgn} \Lambda^0{}_0$ are both continuous functions of the matrix elements $\Lambda^\mu{}_\nu$. Furthermore, $\det \Lambda = \pm 1$ and $\Lambda^0{}_0 \geqslant 1$ or $\leqslant -1$. [The first follows if one takes the determinant of (1–5); the second becomes evident if one looks at the 00 element of (1–5). It reads

$$(\Lambda^0{}_0)^2 - \sum_{j=1}^{3} (\Lambda^j{}_0)^2 = 1.$$

Therefore, $|\Lambda^0{}_0| \geqslant 1$.] Thus, $\det \Lambda$ and $\operatorname{sgn} \Lambda^0{}_0$ must be constant on any one component. The four possibilities are

$$\begin{aligned}
L^\uparrow_+&: \det \Lambda = +1,\ \operatorname{sgn} \Lambda^0{}_0 = +1 \text{ which contains } 1\\
L^\uparrow_-&: \det \Lambda = -1,\ \operatorname{sgn} \Lambda^0{}_0 = +1 \text{ which contains } I_s\\
L^\downarrow_+&: \det \Lambda = +1,\ \operatorname{sgn} \Lambda^0{}_0 = -1 \text{ which contains } I_{st}\\
L^\downarrow_-&: \det \Lambda = -1,\ \operatorname{sgn} \Lambda^0{}_0 = -1 \text{ which contains } I_t\,.
\end{aligned} \tag{1–7}$$

Here, the Lorentz transformations I_s (*space inversion*), I_t (*time inversion*), and I_{st} (*space-time inversion*) are defined by

$$\begin{aligned}
(I_s x)^0 &= x^0 & (I_s x)^j &= -x^j,\ j = 1,2,3\\
(I_t x)^0 &= -x^0 & (I_t x)^j &= x^j,\ j = 1,2,3\\
(I_{st} x) &= -x = (I_s I_t x).
\end{aligned} \tag{1–8}$$

Clearly I_s maps $L^\uparrow_-$ one to one onto $L^\uparrow_+$, I_t maps $L^\downarrow_-$ one to one onto $L^\uparrow_+$, and I_{st} maps $L^\downarrow_+$ one to one onto $L^\uparrow_+$. All Λ for which $\Lambda^0{}_0 \geqslant +1$ are called *orthochronous*, Λ for which $\det \Lambda = +1$ *proper*, and Λ for which $\operatorname{sgn} \Lambda^0{}_0 \det \Lambda = +1$ *orthochorous*. To complete the proof of our assertion it has to be shown that $L^\uparrow_+$ is connected. This is customarily done by proving that any $\Lambda \in L^\uparrow_+$ has a decomposition,

$$\Lambda = \Lambda_1 \Lambda_2 \Lambda_3, \tag{1–9}$$

where Λ_1 and Λ_3 are rotations and Λ_2 is a pure Lorentz transformation along the three-axis, defined by

$$x \to \hat{x} = \Lambda_2 x,$$

where

$$\begin{aligned} \hat{x}^0 &= x^0 \cosh\chi + x^3 \sinh\chi \\ \hat{x}^3 &= x^0 \sinh\chi + x^3 \cosh\chi \\ \hat{x}^1 &= x^1, \qquad \hat{x}^2 = x^2. \end{aligned} \qquad \tanh\chi = \frac{v}{c} \tag{1–10}$$

We can then get from any one transformation of the form (1–9) to any other by varying the axes and angles of rotation of Λ_1 and Λ_3 and the parameter χ of Λ_2 continuously. We shall not prove (1–9) here but refer the reader to Ref. 7 of the bibliography of this chapter. This completes the proof that L has just four components. They are displayed in Figure 1–1.

In Figure 1–1 we have also indicated three important subgroups of L:

$L^\uparrow = L_+^\uparrow \cup L_-^\uparrow$ the *orthochronous Lorentz group*
$L_+ = L_+^\uparrow \cup L_+^\downarrow$ the *proper Lorentz group*
$L_0 = L_+^\uparrow \cup L_-^\downarrow$ the *orthochorous Lorentz group.*

Associated with the restricted Lorentz group, $L_+^\uparrow$, is the group of 2×2 complex matrices of determinant one, which we shall denote $SL(2, C)$. (S stands for special meaning determinant one, L for linear, 2 is the dimension, and C stands for complex.) It is important in describing the transformation properties of spinors. The connection between $L_+^\uparrow$ and $SL(2, C)$ is obtained in the following standard fashion.

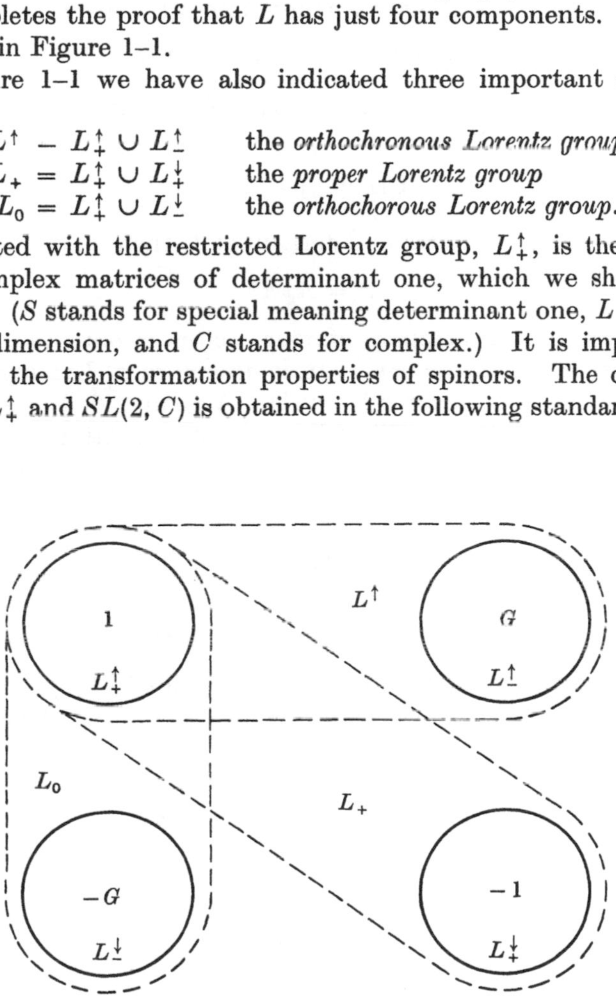

FIGURE 1–1. Connectivity properties of the Lorentz group, L, and its subgroups: the proper Lorentz group, L_+; the orthochronous Lorentz, $L^\uparrow$; the orthochorous Lorentz group, L_0; and the restricted Lorentz group, $L_+^\uparrow$.

If x is any four-vector there is an associated 2×2 matrix given by

$$\underset{\sim}{x} = \begin{pmatrix} x^0 + x^3 & x^1 - ix^2 \\ x^1 + ix^2 & x^0 - x^3 \end{pmatrix} = \sum_{\mu=0}^{3} x^\mu \tau^\mu = x^0 \mathbf{1} + \mathbf{x} \cdot \boldsymbol{\tau}, \tag{1–11}$$

where

$$\tau^0 = \begin{pmatrix} 1 & 0 \\ 0 & 1 \end{pmatrix}, \quad \tau^1 = \begin{pmatrix} 0 & 1 \\ 1 & 0 \end{pmatrix}, \quad \tau^2 = \begin{pmatrix} 0 & -i \\ i & 0 \end{pmatrix}, \quad \tau^3 = \begin{pmatrix} 1 & 0 \\ 0 & -1 \end{pmatrix}.$$

Conversely, every 2×2 matrix X determines a four-vector via

$$x^\mu = \tfrac{1}{2} \operatorname{tr} (X\tau^\mu) \qquad \text{and} \qquad X = \underset{\sim}{x}. \tag{1–12}$$

When x^μ is real, $\underset{\sim}{x}$ is hermitian: $\underset{\sim}{x}^* = \underset{\sim}{x}$; if X is hermitian, (1–12) yields a real four-vector. Note that

$$\det \underset{\sim}{x} = x^\mu x_\mu, \qquad \tfrac{1}{2}[\det (\underset{\sim}{x} + \underset{\sim}{y}) - \det \underset{\sim}{x} - \det \underset{\sim}{y}] = x^\mu y_\mu, \tag{1–13}$$

so if A is any 2×2 matrix of determinant 1, then

$$\hat{\underset{\sim}{x}} = A\underset{\sim}{x}A^* \tag{1–14}$$

defines a real linear mapping of four-vectors x onto four-vectors $\hat{x}$ which satisfies $(\hat{x})^\mu(\hat{y})_\mu = x^\mu y_\mu$. It is therefore a Lorentz transformation, which will be denoted by $\Lambda(A)$. Actually, $\Lambda(A)$ is a restricted Lorentz transformation. This follows from a continuity argument: in (1–14) we can vary A continuously until it is the identity. The corresponding $\Lambda(A)$ varies continuously to reach the identity. Therefore, $\Lambda(A) \in L_+^\uparrow$. Clearly, $\Lambda(-A) = \Lambda(A)$. It is left to the reader to verify that if $\Lambda(A) = \Lambda(B)$ then $A = \pm B$. The correspondence $A \to \Lambda(A)$ has the properties

$$\Lambda(A)\Lambda(B) = \Lambda(AB), \qquad \Lambda(1) = 1. \tag{1–15}$$

In other words, $A \to \Lambda(A)$ is a homomorphism of $SL(2, C)$ onto $L_+^\uparrow$.

We shall have to make use later of another correspondence between vectors and 2×2 matrices

$$\tilde{\underset{\sim}{x}} = \sum_{\mu=0}^{3} x_\mu \tau^\mu = x^0 \mathbf{1} - \mathbf{x} \cdot \boldsymbol{\tau}. \tag{1–16}$$

Note that

$$A^* \tilde{\underset{\sim}{x}} A = \tilde{\underset{\sim}{y}} \qquad \text{where } y = \Lambda(A^{-1})x, \tag{1–17}$$

because

$$(\tau^2)(\tau^\mu)^T(\tau^2)^{-1} = g^{\mu\mu}\tau^\mu,$$

with τ^2 given by (1–11). This implies

$$\tau^2 (\underset{\sim}{x})^T (\tau^2)^{-1} = \tilde{\underset{\sim}{x}}$$

and

$$\tau^2 A^T (\tau^2)^{-1} = A^{-1} \tag{1-18}$$

for A of determinant 1.

Another group associated with the Lorentz group, L, is the *complex Lorentz group*, which we shall denote by $L(C)$. It is essential in the proof of the PCT theorem as we shall see. It is composed of all complex matrices satisfying (1–5). The argument given before to show that $\det \Lambda = \pm 1$ is also valid for the complex Lorentz group $L(C)$, which therefore has at least two components, $L_\pm(C)$, according to the sign of $\det \Lambda$. Actually, it has just two connected components. The transformations 1 and -1, which are disconnected in L are connected in $L(C)$. Consider for example the continuous curve of $\Lambda^\mu{}_\nu$:

$$\Lambda_t = \begin{Bmatrix} \cosh it & 0 & 0 & \sinh it \\ 0 & \cos t & -\sin t & 0 \\ 0 & \sin t & \cos t & 0 \\ \sinh it & 0 & 0 & \cosh it \end{Bmatrix}. \qquad 0 \leqslant t \leqslant \pi.$$

As t varies from 0 to π, this curve connects 1 with -1. The connectivity properties of $L(C)$ are illustrated in Figure 1–2.

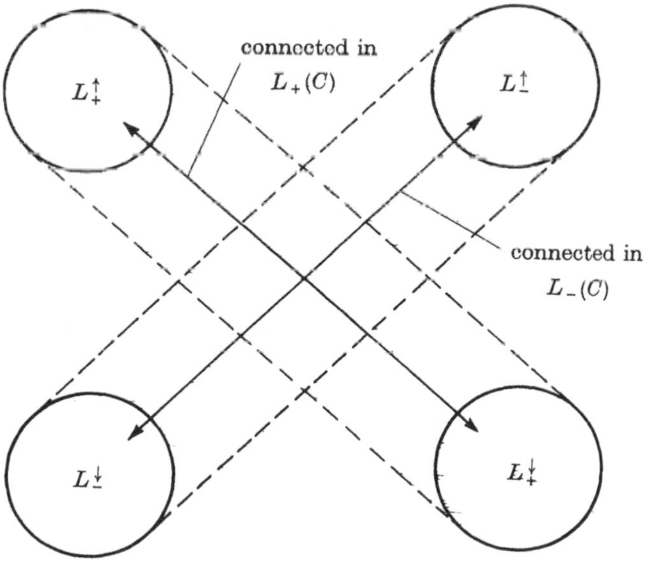

FIGURE 1–2. Connectivity properties of the complex Lorentz group, $L(C)$. There are two connected components: $L_+(C)$, the proper complex Lorentz group, which contains $L_+^\uparrow$ and $L_+^\downarrow$; and $L_-(C)$, which contains $L_-^\uparrow$ and $L_-^\downarrow$.

Just as the restricted Lorentz group, $L_+^\uparrow$, is associated with $SL(2,C)$, the proper complex Lorentz group is associated with $SL(2,C) \otimes SL(2,C)$. This latter group is the set of all pairs of 2×2 matrices of determinant one with the multiplication law

$$\{A_1, B_1\}\{A_2, B_2\} = \{A_1A_2, B_1B_2\}.$$

The connection between the matrix pair $\{A, B\}$ and the corresponding complex Lorentz transformation $\Lambda(A, B)$ is given by the analogue of (1–14).

$$\hat{\underset{\sim}{x}} = A\underset{\sim}{x}B^T. \tag{1–19}$$

Here $\underset{\sim}{x}$ is again defined by (1–11), but the vector x^μ is complex. Clearly

$$\Lambda(A_1, B_1)\Lambda(A_2, B_2) = \Lambda(A_1A_2, B_1B_2) \tag{1–20}$$

and $\Lambda(1, 1) = 1$. It is easy to see that the only matrix pairs which yield a given $\Lambda(A, B)$ are $(\pm A, \pm B)$. In particular,

$$\Lambda(-1, 1) = \Lambda(1, -1) = -1. \tag{1–21}$$

To each of the groups we have considered so far there is a corresponding inhomogeneous group, whose elements are pairs consisting of a translation and a homogeneous transformation. For example, the *Poincaré group*, $\mathscr{P}$, has elements $\{a, \Lambda\}$, where $\Lambda \in L$, with the multiplication law

$$\{a_1, \Lambda_1\}\{a_2, \Lambda_2\} = \{a_1 + \Lambda_1a_2, \Lambda_1\Lambda_2\}, \tag{1–22}$$

as follows from $\{a, \Lambda\}$'s interpretation as a linear transformation

$$x \to \Lambda x + a.$$

$\mathscr{P}$ has four components distinguished by det Λ and sgn $\Lambda^0{}_0$, namely, $\mathscr{P}_+^\uparrow$, $\mathscr{P}_-^\uparrow$, $\mathscr{P}_+^\downarrow$, and $\mathscr{P}_-^\downarrow$. The *complex Poincaré group*, $\mathscr{P}(C)$ admits complex translations but also the multiplication law (1–22). It has two components $\mathscr{P}_\pm(C)$ which are distinguished by det Λ. The inhomogeneous group corresponding to $SL(2,C)$ has no special name; we shall call it the *inhomogeneous* $SL(2,C)$. It consists of pairs $\{a, A\}$, where a is a translation and $A \in SL(2,C)$, and a multiplication law

$$\{a_1, A_1\}\{a_2, A_2\} = \{a_1 + \Lambda(A_1)a_2, A_1A_2\}. \tag{1–23}$$

Sometimes, we shall find it convenient to use the obvious shorthand Aa for $\Lambda(A)a$ and we shall do so in the following. Analogously, the inhomogeneous group of $SL(2,C) \otimes SL(2,C)$ is defined as the set of pairs $\{a, \{A, B\}\}$, where a is a complex translation and $A, B \in SL(2,C)$, and a multiplication law

$$\{a_1, \{A_1, B_1\}\}\{a_2, \{A_2, B_2\}\} = \{a_1 + \Lambda(A_1, B_1)a_2, \{A_1A_2, B_1B_2\}\}. \tag{1–24}$$

In all the following, all the theories considered will be invariant under $\mathscr{P}_+^\uparrow$ or some other subgroup of $\mathscr{P}$. The corresponding unitary or anti-unitary transformations of states will be denoted $U(a, \Lambda)$.

This completes our discussion of the groups themselves. We need in addition some information on the matrix representations of $SL(2,C)$. This is the basic raw material out of which the transformation law of fields will be constructed in Chapter 3.† We recall that an arbitrary matrix representation of $SL(2,C)$, i.e., a correspondence $A \to S(A)$ such that $S(\mathbf{1}) = 1$ and $S(A)S(B) = S(AB)$, can be written in the form

$$T\begin{Bmatrix} S_1(A) & 0 & 0 & \dots \\ 0 & S_2(A) & 0 & \dots \\ \vdots & \vdots & & \ddots \end{Bmatrix}T^{-1}, \tag{1–25}$$

where the $A \to S_1(A)$, $A \to S_2(A)$, etc., are irreducible and T is some fixed linear transformation. Thus, we need only describe the irreducible representations.

Consider a set of quantities $\xi_{\alpha_1\ldots\alpha_j\dot\beta_1\ldots\dot\beta_k}$, where the α's and $\dot\beta$'s take the values 1 and 2, and ξ is symmetric under permutations of the α's and also under permutations of the $\dot\beta$'s. For each $A \in SL(2,C)$ we define a linear transformation of the ξ's according to

$$\xi_{\alpha_1\ldots\alpha_j\dot\beta_1\ldots\dot\beta_k} \to \sum_{(\rho)\,(\dot\sigma)} A_{\alpha_1\rho_1}\ldots A_{\alpha_j\rho_j}\bar{A}_{\dot\beta_1\dot\sigma_1}\ldots\bar{A}_{\dot\beta_k\dot\sigma_k}\xi_{\rho_1\ldots\rho_j\dot\sigma_1\ldots\dot\sigma_k}. \tag{1–26}$$

[The dot over the index simply means that this index transforms according to $\bar{A}$ instead of A; the symbol (ρ) stands for $\rho_1\ldots\rho_j$; the symbol $(\dot\sigma)$ for $\dot\sigma_1\ldots\dot\sigma_k$.] This representation of $SL(2,C)$ is usually denoted $\mathscr{D}^{(j/2,k/2)}$. Every irreducible representation is equivalent to one of these.

If the A are restricted to be unitary as well as of determinant one we get a subgroup SU_2 of $SL(2,C)$. Clearly, SU_2 corresponds via (1–14) to the rotation group of three-space. [The trace of (1–14) reads $2\hat{x}^0 = 2x^0$ if A is unitary.] The irreducible representations of SU_2 are

† The reader who is unfamiliar with the finite-dimensional representations of $SL(2,C)$ might do well to skip the rest of this section on first reading. Readable accounts of these representations are to be found in Refs. 9 and 10. For the complex Lorentz group see also Ref. 7. We have included this material here because we want to treat an arbitrary spinor field in Chapters 3 and 4.

equivalent to one of the $A \to \mathscr{D}^{(j/2,0)}(A)$ with $A \in SU_2$; these are usually denoted $\mathscr{D}^{(j/2)}$ and are just the usual transformations associated with a system of angular momentum $j/2$. The representation of SU_2 obtained by restricting $A \to \mathscr{D}^{(j/2,k/2)}(A)$ to $A \in SU_2$ is not irreducible. In fact, it is just the direct product of $\mathscr{D}^{(j/2)}$ with $\mathscr{D}^{(k/2)}$ so, by the usual laws of combination of angular momentum, it is a sum of

$$\mathscr{D}^{\left(\frac{j+k}{2}\right)}, \mathscr{D}^{\left(\frac{j+k}{2}-1\right)}, \ldots \mathscr{D}^{\left|\frac{j-k}{2}\right|}.$$

The representation $\mathscr{D}^{(j/2,k/2)}$ can be continued analytically from $SL(2,C)$ to $SL(2,C) \otimes SL(2,C)$ if one replaces $\{A,\bar{A}\}$ by $\{A,B\}$. We define the matrix $\mathscr{D}^{(j/2,k/2)}(A,B)$ by the linear correspondence

$$\xi_{\alpha_1\ldots\alpha_j\beta_1\ldots\beta_k} \to \sum_{(\rho)\,(\dot{\sigma})} A_{\alpha_1\rho_1}\ldots A_{\alpha_j\rho_j} B_{\beta_1\dot{\sigma}_1}\ldots B_{\beta_k\dot{\sigma}_k} \xi_{\rho_1\ldots\rho_j\dot{\sigma}_1\ldots\dot{\sigma}_k}.$$

Clearly,

$$\begin{aligned} \mathscr{D}^{\left(\frac{j}{2},\frac{k}{2}\right)}(1,-1) &= (-1)^k \\ \mathscr{D}^{\left(\frac{j}{2},\frac{k}{2}\right)}(-1,1) &= (-1)^j, \end{aligned} \qquad (1\text{–}27)$$

a relation we shall need in our discussion of the PCT theorem.

The spinor fields we consider will sometimes also have definite transformation laws under space inversion I_s (or P as it is also called), time inversion I_t (or T), and charge conjugation C. The product of these transformations in one order or another, say PCT, is always a symmetry in local field theory whether or not the individual factors are; this is the famous PCT theorem which we shall prove in Chapter 4. To justify the name PCT we shall here display P, C, and T separately for a general spinor quantity.

For clarity, we want to recall the relation between our description of symmetries and that which was standard in the quantum field theory of 30 years ago. Then a symmetry was given as a substitution law, $A \to \hat{A}$, for the operators A in terms of which the equations of the theory were formulated. For example, in the case under discussion in (1–3), $\mathbf{q}_i(t) \to \hat{\mathbf{q}}_i(t) = -\mathbf{q}_i(t)$, $\mathbf{p}_i(t) \to \hat{\mathbf{p}}_i(t) = -\mathbf{p}_i(t)$ is space inversion. Symmetry was guaranteed by requiring that the equations of motion be form-invariant under the substitution $A \to \hat{A}$. It was tacitly assumed that to each substitution there exists a unitary transformation $\Phi \to \hat{\Phi}$ such that

$$(\Phi, \hat{A}\Phi) = (\hat{\Phi}, A\hat{\Phi}). \qquad (1\text{–}28)$$

This equation holds for all Φ, so if $\hat{\Phi} = U\Phi$, it is equivalent to

$$\hat{A} = U^{-1}AU. \qquad (1\text{–}29)$$

However, it soon became clear that there are also important symmetries for which $\Phi \to \hat{\Phi}$ is an anti-unitary correspondence. For these

$$(\hat{\Phi},A\hat{\Phi}) = (U\Phi,AU\Phi) = \overline{(\Phi,U^{-1}AU\Phi)} = (\Phi,(U^{-1}AU)^*\Phi),$$

and thus, instead of (1–29), we have

$$\hat{A} = (U^{-1}AU)^* \tag{1–30}$$

if U is anti-unitary. Equation (1–30) implies

$$\widehat{AB} = (U^{-1}ABU)^* = (U^{-1}BU)^*(U^{-1}AU)^* = \hat{B}\hat{A}, \tag{1–31}$$

which gives rise to rather curious substitution rules. For example, the PCT symmetry Θ has the property that, for a charged scalar field φ (a notion that will be defined precisely in Chapter 3),

$$\Theta^{-1}\varphi(x)\Theta = \varphi(-x)^*, \tag{1–32}$$

where Θ is anti-unitary. Thus the substitution rule for PCT is: replace $\varphi(x)$ by $\varphi(-x)$ and write all operator products in reverse order. Of late it has become customary to formulate symmetries in a more direct physical way, as we have done in Section 1–2. However, in the next few paragraphs it is useful to admit both formulations, since this helps in deciding on the transformation laws of general spinor fields.

We begin by considering two simple cases, the coordinate vector x and a Dirac spinor ψ.

To each real vector x in space-time we have associated a 2×2 hermitian matrix $\underset{\sim}{x}$. It is conventional to take the matrix elements of $\underset{\sim}{x}$ as the components of a second-rank spinor with one dotted and one undotted index,

$$x_{\alpha\dot{\beta}} = (\underset{\sim}{x})_{\alpha\beta}. \tag{1–33}$$

This is consistent with the transformation law $\underset{\sim}{x} \to A\underset{\sim}{x}A^*$ and our convention that dotted indices should transform with $\bar{A}$. Now the operations P and T are given respectively by

$$P\colon \quad \underset{\sim}{x} \to \zeta\bar{\underset{\sim}{x}}\zeta^{-1} \qquad \text{where } \zeta = i\tau^2 \tag{1–34}$$

and

$$T\colon \quad \underset{\sim}{x} \to -\zeta\bar{\underset{\sim}{x}}\zeta^{-1}. \tag{1–35}$$

Here $\bar{\underset{\sim}{x}}$ is the complex conjugate of $\underset{\sim}{x}$.

In order to get (1–34) and (1–35) into the form of linear transformations we have to introduce the conjugate spinor

$$x_{\dot{\alpha}\beta} = (\bar{\underset{\sim}{x}})_{\dot{\alpha}\beta}. \tag{1–36}$$

Then, the hermiticity of $\underset{\approx}{x}$ is expressed by $x_{\dot{\alpha}\beta} = x_{\beta\dot{\alpha}}$. The pair transforms under inversions according to

$$P: \quad \begin{pmatrix} x_{\alpha\dot{\beta}} \\ x_{\dot{\alpha}\beta} \end{pmatrix} \to \begin{pmatrix} 0 & \zeta \otimes \zeta \\ \zeta \otimes \zeta & 0 \end{pmatrix} \begin{pmatrix} x_{\alpha\dot{\beta}} \\ x_{\dot{\alpha}\beta} \end{pmatrix} \tag{1–37}$$

and

$$T: \quad \begin{pmatrix} x_{\alpha\dot{\beta}} \\ x_{\dot{\alpha}\beta} \end{pmatrix} \to \begin{pmatrix} 0 & -\zeta \otimes \zeta \\ -\zeta \otimes \zeta & 0 \end{pmatrix} \begin{pmatrix} x_{\alpha\dot{\beta}} \\ x_{\dot{\alpha}\beta} \end{pmatrix}. \tag{1–38}$$

Here $\zeta \otimes \zeta$ is understood to be the linear transformation in which the first ζ acts on the first index and the second on the second.

Up to this point, the discussion has dealt with a complex-valued second-rank spinor. Now we pass to the case of a set of operators $\xi_{\alpha\dot{\beta}}$ in quantum theory which has the same transformation law. (We need not commit ourselves to any particular physical interpretation.) Equations (1–37) and (1–38) are then to be interpreted as substitution rules, with $(\xi_{\alpha\dot{\beta}})^* = \xi_{\dot{\alpha}\beta}$. Because space inversion is represented by a unitary operator $U(I_s)$, and time inversion by an anti-unitary operator $U(I_t)$, we have

$$P: \quad U(I_s)^{-1} \begin{pmatrix} \xi_{\alpha\dot{\beta}} \\ \xi_{\dot{\alpha}\beta} \end{pmatrix} U(I_s) = \begin{pmatrix} 0 & \zeta \otimes \zeta \\ \zeta \otimes \zeta & 0 \end{pmatrix} \begin{pmatrix} \xi_{\alpha\dot{\beta}} \\ \xi_{\dot{\alpha}\beta} \end{pmatrix}, \tag{1–39}$$

but

$$T: \quad \left[U(I_t)^{-1} \begin{pmatrix} \xi_{\alpha\dot{\beta}} \\ \xi_{\dot{\alpha}\beta} \end{pmatrix} U(I_t) \right]^* = -\begin{pmatrix} 0 & \zeta \otimes \zeta \\ \zeta \otimes \zeta & 0 \end{pmatrix} \begin{pmatrix} \xi_{\alpha\dot{\beta}} \\ \xi_{\dot{\alpha}\beta} \end{pmatrix},$$

which is

$$U(I_t)^{-1} \begin{pmatrix} \xi_{\alpha\dot{\beta}} \\ \xi_{\dot{\alpha}\beta} \end{pmatrix} U(I_t) = -\begin{pmatrix} \zeta \otimes \zeta & 0 \\ 0 & \zeta \otimes \zeta \end{pmatrix} \begin{pmatrix} \xi_{\alpha\dot{\beta}} \\ \xi_{\dot{\alpha}\beta} \end{pmatrix}. \tag{1–40}$$

The second simple example is a four-component Dirac spinor ψ. Here one can get another non-trivial illustration of the P and T transformation and a C transformation as well. ψ satisfies

$$\left(\gamma^\mu \frac{\partial}{\partial x^\mu} + m\right)\psi(x) = 0, \tag{1–41}$$

where the 4×4 matrices γ^μ are solutions of

$$\gamma^\mu\gamma^\nu + \gamma^\nu\gamma^\mu = -2g^{\mu\nu}. \tag{1–42}$$

Then there is a 4×4 representation of $SL(2,C)$, $A \to S(A)$ which satisfies

$$S(A)^{-1}\gamma^\mu S(A) = \Lambda^\mu{}_\nu(A)\gamma^\nu. \tag{1–43}$$

The spinor transforms as $\psi(x) \to S(A)\psi(A^{-1}x)$. The relation between A and $S(A)$ can be made explicit:

$$S(A) = (a^0\mathbf{1} + \mathbf{a}\cdot\boldsymbol{\sigma})\tfrac{1}{2}(1 + i\gamma^5) + (\bar{a}^0\mathbf{1} - \bar{\mathbf{a}}\cdot\boldsymbol{\sigma})\tfrac{1}{2}(1 - i\gamma^5), \quad (1\text{–}44)$$

where

$$\gamma^5 = \gamma^0\gamma^1\gamma^2\gamma^3 \qquad \text{and} \qquad \boldsymbol{\sigma} = -i\gamma^5\boldsymbol{\alpha} = -i\gamma^5\gamma^0\boldsymbol{\gamma}$$

and

$$A = \sum_{\mu=0}^{3} a^\mu\tau^\mu, \qquad \det A = a^2 = 1.$$

The proof that (1–44) actually satisfies (1–43) requires a certain amount of γ-gymnastics, and makes essential use of the relation (1–18).

A possible choice of substitution rules corresponding to P, C, and T for the Dirac equation is†

$$\begin{aligned} P:&\quad \psi(x) \to -\gamma^0\psi(I_s x) \\ C:&\quad \psi(x) \to C^{-1}\overline{\psi(x)} = \psi^C(x) \\ T:&\quad \psi(x) \to -\gamma^0\gamma^5\psi^C(I_t x) = -\gamma^0\gamma^5 C^{-1}\overline{\psi(I_t x)}, \end{aligned} \quad (1\text{–}45)$$

where C is a matrix satisfying

$$C\gamma^\mu C^{-1} = \bar{\gamma}^\mu.$$

A permissible choice of the γ's and C is

$$\gamma^0 = -i\begin{pmatrix} 0 & 1 \\ 1 & 0 \end{pmatrix}, \qquad \boldsymbol{\gamma} = -i\begin{pmatrix} 0 & -\boldsymbol{\tau} \\ \boldsymbol{\tau} & 0 \end{pmatrix}, \qquad C = -\begin{pmatrix} 0 & -\zeta \\ \zeta & 0 \end{pmatrix}$$

which gives

$$i\gamma^5 = \begin{pmatrix} 1 & 0 \\ 0 & -1 \end{pmatrix}, \qquad \boldsymbol{\sigma} = \begin{pmatrix} \boldsymbol{\tau} & 0 \\ 0 & \boldsymbol{\tau} \end{pmatrix}, \qquad S(A) = \begin{pmatrix} A & 0 \\ 0 & \zeta^{-1}\bar{A}\zeta \end{pmatrix}.$$

In this case the Dirac equation splits up into a pair of equations for the two-component objects ξ,χ, where $\psi = \begin{pmatrix} \xi \\ \chi \end{pmatrix}$:

$$\begin{aligned} i\left(\frac{-\partial}{\partial x^0} + \sum_{j=1}^{3} \tau^j \frac{\partial}{\partial x^j}\right)\chi + m\xi &= 0 \\ i\left(\frac{-\partial}{\partial x^0} - \sum_{j=1}^{3} \tau^j \frac{\partial}{\partial x^j}\right)\xi + m\chi &= 0. \end{aligned} \quad (1\text{–}46)$$

The two-component quantity $\eta_{\dot{\alpha}} = (\zeta\chi)_{\dot{\alpha}}$ transforms in the same way

† Here, $\overline{\psi(x)}$ denotes the complex conjugate of $\psi(x)$ and not, as is common, $\overline{\psi(x)}^T\gamma^0$.

under $SL(2,C)$ as the complex conjugate of ξ (although it is not equal to it). In terms of $\xi_\alpha, \eta_{\dot\alpha}$ the substitution rules become

$$
\begin{aligned}
P:&\quad \begin{pmatrix}\xi_\alpha\\ \eta_{\dot\alpha}\end{pmatrix} \to i\begin{pmatrix}0 & -\zeta\\ \zeta & 0\end{pmatrix}\begin{pmatrix}\xi_\alpha\\ \eta_{\dot\alpha}\end{pmatrix}\\
C:&\quad \begin{pmatrix}\xi_\alpha\\ \eta_{\dot\alpha}\end{pmatrix} \to \begin{pmatrix}0 & 1\\ 1 & 0\end{pmatrix}\begin{pmatrix}\bar\xi_\alpha\\ \bar\eta_{\dot\alpha}\end{pmatrix} \qquad (1\text{–}47)\\
T:&\quad \begin{pmatrix}\xi_\alpha\\ \eta_{\dot\alpha}\end{pmatrix} \to \begin{pmatrix}\zeta & 0\\ 0 & \zeta\end{pmatrix}\begin{pmatrix}\bar\xi_\alpha\\ \bar\eta_{\dot\alpha}\end{pmatrix}.
\end{aligned}
$$

We now interpret these as substitution rules for operators according to (1–29) for P and C, and (1–30) for T. Thus

$$
\begin{aligned}
U(I_s)^{-1}\begin{pmatrix}\xi_\alpha\\ \eta_{\dot\alpha}\end{pmatrix}U(I_s) &= i\begin{pmatrix}0 & -\zeta\\ \zeta & 0\end{pmatrix}\begin{pmatrix}\xi_\alpha\\ \eta_{\dot\alpha}\end{pmatrix}\\
U(C)^{-1}\begin{pmatrix}\xi_\alpha\\ \eta_{\dot\alpha}\end{pmatrix}U(C) &= \begin{pmatrix}0 & 1\\ 1 & 0\end{pmatrix}\begin{pmatrix}\xi^*_\alpha\\ \eta^*_{\dot\alpha}\end{pmatrix} \qquad (1\text{–}48)\\
U(I_t)^{-1}\begin{pmatrix}\xi_\alpha\\ \eta_{\dot\alpha}\end{pmatrix}U(I_t) &= \begin{pmatrix}\zeta & 0\\ 0 & \zeta\end{pmatrix}\begin{pmatrix}\xi_\alpha\\ \eta_{\dot\alpha}\end{pmatrix}
\end{aligned}
$$

$$
\Theta^{-1}\begin{pmatrix}\xi_\alpha\\ \eta_{\dot\alpha}\end{pmatrix}\Theta = i\begin{pmatrix}-1 & 0\\ 0 & 1\end{pmatrix}\begin{pmatrix}\xi^*_\alpha\\ \eta^*_{\dot\alpha}\end{pmatrix}, \qquad (1\text{–}49)
$$

where $U(C)$ is a unitary operator which takes a state to its charge conjugate state, and Θ is the anti-unitary PCT operator $U(I_s)U(C)U(I_t)$. Comparison of the two examples suggests the following definitions for a general spinor:

Substitution rules:

$$
\begin{aligned}
P:&\quad \begin{pmatrix}\xi_{(\alpha)(\dot\beta)}\\ \eta_{(\dot\alpha)(\beta)}\end{pmatrix} \to \begin{pmatrix} i & j+k \text{ odd}\\ 1 & j+k \text{ even}\end{pmatrix}\\
&\qquad \times \begin{pmatrix} 0 & (-1)^j\zeta\otimes\zeta\otimes\cdots\\ (-1)^k\zeta\otimes\zeta\otimes\cdots & 0\end{pmatrix}\begin{pmatrix}\xi_{(\alpha)(\dot\beta)}\\ \eta_{(\dot\alpha)(\beta)}\end{pmatrix}\\
C:&\quad \begin{pmatrix}\xi_{(\alpha)(\dot\beta)}\\ \eta_{(\dot\alpha)(\beta)}\end{pmatrix} \to \begin{pmatrix}0 & 1\\ 1 & 0\end{pmatrix}\begin{pmatrix}\bar\xi_{(\alpha)(\dot\beta)}\\ \bar\eta_{(\dot\alpha)(\beta)}\end{pmatrix} \qquad (1\text{–}50)\\
T:&\quad \begin{pmatrix}\xi_{(\alpha)(\dot\beta)}\\ \eta_{(\dot\alpha)(\beta)}\end{pmatrix} \to \begin{pmatrix}\zeta\otimes\zeta\otimes\cdots & 0\\ 0 & \zeta\otimes\zeta\otimes\cdots\end{pmatrix}\begin{pmatrix}\bar\xi_{(\alpha)(\dot\beta)}\\ \bar\eta_{(\dot\alpha)(\beta)}\end{pmatrix}
\end{aligned}
$$

$$PCT: \begin{pmatrix} \xi_{(\alpha)(\beta)} \\ \eta_{(\dot\alpha)(\beta)} \end{pmatrix} \to \begin{pmatrix} -i, & j+k \text{ odd} \\ 1, & j+k \text{ even} \end{pmatrix} \begin{pmatrix} (-1)^j & 0 \\ 0 & (-1)^k \end{pmatrix} \begin{pmatrix} \xi_{(\alpha)(\beta)} \\ \eta_{(\dot\alpha)(\beta)} \end{pmatrix}. \tag{1–51}$$

In these equations $\xi_{(\alpha)(\beta)} = \xi_{\alpha_1\ldots\alpha_j\beta_1\ldots\beta_k}$, $\eta_{(\dot\alpha)(\beta)} = \eta_{\dot\alpha_1\ldots\dot\alpha_j\beta_1\ldots\beta_k}$.

In terms of transformations of the states, we have

$$U(I_s)^{-1} \begin{pmatrix} \xi_{(\alpha)(\beta)} \\ \eta_{(\dot\alpha)(\beta)} \end{pmatrix} U(I_s) =$$

$$\begin{pmatrix} i, & j+k \text{ odd} \\ 1, & j+k \text{ even} \end{pmatrix} \begin{pmatrix} 0 & (-1)^j \zeta \otimes \zeta \otimes \cdots \\ (-1)^k \zeta \otimes \zeta \otimes \cdots & 0 \end{pmatrix} \begin{pmatrix} \xi_{(\alpha)(\beta)} \\ \eta_{(\dot\alpha)(\beta)} \end{pmatrix}$$

$$U(C)^{-1} \begin{pmatrix} \xi_{(\alpha)(\beta)} \\ \eta_{(\dot\alpha)(\beta)} \end{pmatrix} U(C) = \begin{pmatrix} 0 & 1 \\ 1 & 0 \end{pmatrix} \begin{pmatrix} \xi^*_{(\dot\alpha)(\beta)} \\ \eta^*_{(\dot\alpha)(\beta)} \end{pmatrix} \tag{1–52}$$

$$U(I_t)^{-1} \begin{pmatrix} \xi_{(\alpha)(\beta)} \\ \eta_{(\dot\alpha)(\beta)} \end{pmatrix} U(I_t) = \begin{pmatrix} \zeta \otimes \zeta \otimes \cdots & 0 \\ 0 & \zeta \otimes \zeta \otimes \cdots \end{pmatrix} \begin{pmatrix} \xi_{(\alpha)(\beta)} \\ \eta_{(\dot\alpha)(\beta)} \end{pmatrix}$$

leading to

$$\Theta^{-1} \begin{pmatrix} \xi_{(\alpha)(\beta)} \\ \eta_{(\dot\alpha)(\beta)} \end{pmatrix} \Theta = \begin{pmatrix} i & j+k \text{ odd} \\ 1 & j+k \text{ even} \end{pmatrix} \begin{pmatrix} (-1)^j & 0 \\ 0 & (-1)^k \end{pmatrix} \begin{pmatrix} \xi^*_{(\dot\alpha)(\beta)} \\ \eta^*_{(\dot\alpha)(\beta)} \end{pmatrix}, \tag{1–53}$$

which we take as our definition of the transformation law of a general spinor under PCT. The particular choices of phases that have been made in (1–50) and (1–52) are highly arbitrary. They differ from (1–37) and (1–38) by a sign. Such distinctions are in general (but not always) physically significant, and give the difference, for example, between vectors and pseudo-vectors. The problem of the physical significance of phases is discussed again in Section 3–5. There are many other laws besides 1–50 and 1–52 which lead to the same PCT transformation (1–53) Our purpose here is merely to give one example.

1–4. RELATIVISTIC TRANSFORMATION LAWS OF STATES

We begin by describing the relativistic transformation laws of some simple and important states and gradually work our way toward a description of behavior of a complete theory under Lorentz transformation.

THE VACUUM STATE

The vacuum state, Ψ_0, looks the same to all observers. It has zero energy, momentum, and angular momentum. Picking a vector, Ψ_0, to represent Ψ_0, we may write as the simplest form of transformation law consistent with these statements:

$$U(a, \Lambda)\Psi_0 = \Psi_0 \qquad \text{for } \{a, \Lambda\} \in \mathscr{P}_+^\uparrow. \tag{1–54}$$

This gives the so-called identity representation of $\mathscr{P}_+^\uparrow$: $\{a, \Lambda\} \to 1$. It is important to note that the vacuum possesses this transformation law quite independently of what interactions there are in other states of the system.

ONE-PARTICLE STATES AND OTHER ELEMENTARY SYSTEMS

Consider first the simplest case of a single scalar particle of mass m, alone in the world without external fields. Its four-momentum satisfies $p^2 = m^2$ with $p_0 \geqslant 0$ and if we describe its states, Ψ, by complex-valued functions, $\Psi(p)$, which satisfy†

$$\int |\Psi(p)|^2 d\Omega_m(p) < \infty,$$

where

$$d\Omega_m(p) = \frac{d^3p}{\sqrt{m^2 + \mathbf{p}^2}}, \tag{1–55}$$

we have as a reasonable transformation law

$$(U(a, \Lambda)\Psi)(p) = e^{ip\cdot a}\Psi(\Lambda^{-1}p). \tag{1–56}$$

This $U(a, \Lambda)$ is unitary because

$$\begin{aligned}(U(a, \Lambda)\Phi, U(a, \Lambda)\Psi) &= \int \overline{(U(a, \Lambda)\Phi)(p)}(U(a, \Lambda)\Psi)(p)\, d\Omega_m(p) \\ &= \int \bar{\Phi}(\Lambda^{-1}p)\Psi(\Lambda^{-1}p)\, d\Omega_m(p) \\ &= \int \bar{\Phi}(p)\Psi(p)\, d\Omega_m(\Lambda p) = (\Phi, \Psi),\end{aligned}$$

since $d\Omega_m(\Lambda p) = d\Omega_m(p)$ because $d\Omega_m$ is the invariant volume on the hyperboloid $p^2 = m^2$.

To describe the transformation law of a particle of mass $m > 0$ and arbitrary spin, s, we need a little more machinery. We describe a

† The relation of this description to the usual Dirac formalism is as follows. Let $|p\rangle$ be the state of four-momentum p (it is not a state in the sense used in this book because it is not normalizable), with the continuum normalization $\langle p' \mid p\rangle = \delta(\mathbf{p} - \mathbf{p}')p^0$. An arbitrary state, Φ, of mass m then can be expanded $\Phi = \int d\Omega_m(p) \mid p\rangle\langle p \mid \Phi\rangle$. The scalar product of two such states is

$$(\Phi, \Psi) = \int d\Omega_m(p)\overline{\langle p \mid \Phi\rangle}\langle p \mid \Psi\rangle.$$

The transformation law of the $|p\rangle$ under $\{a, \Lambda\} \in \mathscr{P}_+^\uparrow$ is

$$U(a, \Lambda) \mid p\rangle = e^{i\Lambda p\cdot a} \mid \Lambda p\rangle.$$

Evidently our $\Psi(p)$ is just $\langle p \mid \Psi\rangle$.

state, Ψ, by a set of complex-valued functions of p labeled by $2s$ spinor indices $\alpha_1 \ldots \alpha_{2s}$:

$$\Psi_{\alpha_1 \ldots \alpha_{2s}}(p)$$

and symmetric under permutations of the α's. The scalar product is

$$(\Phi,\Psi) = \int d\Omega_m(p) \sum_{\alpha_1 \ldots \alpha_{2s}} \sum_{\beta_1 \ldots \beta_{2s}} \bar{\Phi}_{\alpha_1 \ldots \alpha_{2s}}(p) \times \prod_{j=1}^{2s} (\tilde{p}/m)_{\alpha_j \beta_j} \Psi_{\beta_1 \ldots \beta_{2s}}(p), \quad (1\text{–}57)$$

where $\tilde{p} = p_\mu \tau^\mu$. It is perhaps not obvious that this scalar product is positive definite, i.e., that $(\Phi, \Phi) \geqslant 0$ and $\Phi = 0$ if $(\Phi, \Phi) = 0$, but in fact that is the case because the matrix $\tilde{p}/m$ is positive definite. (Its trace and determinant are $2p^0/m$ and 1, respectively, so both its proper values are positive.) We give the transformation law under $\mathscr{P}_+^\uparrow$ by constructing a representation of the inhomogeneous $SL(2,C)$:

$$(U(a, A)\Psi)_{\alpha_1 \ldots \alpha_{2s}}(p) = e^{ip\cdot a} \sum_{\beta_1 \ldots \beta_{2s}} \prod_{j=1}^{2s} (A)_{\alpha_j \beta_j} \Psi_{\beta_1 \ldots \beta_{2s}}(A^{-1}p). \quad (1\text{–}58)$$

That this $U(a, A)$ leaves the scalar product invariant follows from the identity

$$A^*\tilde{p}A = \widetilde{A^{-1}p}. \quad (1\text{–}59)$$

See (1–17) for the proof of (1–59). Note that (1–58) uses only undotted indices. To see that (1–58) describes a system of spin s one can look at the transformations A which are unitary. These leave the vector $p = (m,0,0,0)$ invariant. Clearly, the amplitudes for the particle at rest are transformed under SU_2 according to the representation $\mathscr{D}^{(s)}$ of spin s described after (1–26), and that is precisely what is meant by the statement that the system has spin s.

Analogous but somewhat different transformation laws exist for spin s and mass zero. We refer the reader to the Bibliography for further information.

It is an essential feature of these transformation laws that they provide a maximal commuting set of observables. One can take, for example, the momentum $\mathbf{p}$ and the component of angular momentum along $\mathbf{p}$. The existence of such a maximal commuting set is associated with the fact that the representation is *irreducible*, i.e., any operator which commutes with all $U(a, A)$ is a constant multiple of the identity.

If the transformation law $\{a, A\} \rightarrow U(a, A)$ of the states of a physical system under $\mathscr{P}_+^\uparrow$ is irreducible, we call it an *elementary system*. It is characteristic of an elementary system that there is no way of knowing from the $U(a,\Lambda)$ alone whether it is elementary or composite in the usual sense in which an electron is elementary and a deuteron composite. Its only ascertainable properties are its mass and spin. We emphasize this by introducing the symbol $[m,s]$ to stand for the transformation law (1–58). In a theory of elementary particles, the elementary systems span a Hilbert space $\mathscr{H}_1$, which is a proper subspace of the Hilbert space $\mathscr{H}$ of the theory.

We expect all stable elementary particles to be elementary systems, at least to the extent that we can neglect the special problems arising from particles of zero mass. This proviso is related to the so-called infrared problem. A particle that is coupled to a zero mass field acquires a cloud of virtual particles which can alter its transformation law; it may not have a definite mass. To describe the particle one may have to make a wave packet in mass which describes the behavior of the cloud. These are very interesting and important matters but we want to avoid them in this book, so we shall assume whenever necessary that there are no zero mass particles in the systems under consideration.

TWO OR MORE PARTICLE STATES; COLLISION STATES

Given non-interacting systems, one can get a description of the combined system formed from them as constituents by taking product wave functions. If some of the systems involved are identical, then the product wave functions have to be symmetrized or anti-symmetrized, according to the statistics which the identical systems satisfy. This operation uniquely determines a transformation law of the combined system. For example, two free scalar particles of masses m_1 and m_2, respectively, combined have states Ψ described by complex-valued functions $\Psi(p_1, p_2)$, where $p_1{}^2 = m_1{}^2$, $p_2{}^2 = m_2{}^2$ and $p_1{}^0 \geqslant 0$, $p_2{}^0 \geqslant 0$. The scalar product is

$$(\Phi, \Psi) = \iint d\Omega_{m_1}(p_1)\, d\Omega_{m_2}(p_2)\bar{\Phi}(p_1, p_2)\Psi(p_1, p_2)$$

and the transformation law is

$$(U(a, \Lambda)\Phi)(p_1, p_2) = e^{i(p_1+p_2)\cdot a}\Phi(\Lambda^{-1}p_1, \Lambda^{-1}p_2).$$

In a shorthand notation, this transformation law is $[m_1, 0] \otimes [m_2, 0]$. If $m_1 = m_2 = m$, and the $\Psi(p_1, p_2)$ are symmetrical under permutation of their arguments, the shorthand for the transformation law is $([m, 0] \otimes [m, 0])_S$. If we write $M^2 = (p_1 + p_2)^2$, M is the mass of the combined system and it varies continuously from $m_1 + m_2$ to infinity as the relative momentum of the particles varies in magnitude from zero to infinity. The amplitude for $\mathbf{p}_1 + \mathbf{p}_2 = 0$ still depends on the relative momentum so, under rotations, one can have all orbital angular momenta. That means that the combined system contains subsystems $[M,\ell]$ of every integer spin, ℓ, for each mass, M, in a continuum extending from $m_1 + m_2$ to infinity. Moreover, for two-particle systems, the $[M,\ell]$ uniquely define the state, i.e., the infinitesimal translation and Lorentz transformation operators $P_\mu, M^{\mu\nu}$ form an irreducible set of operators. Thus, each representation enters with multiplicity one.

Now we ask what happens to this transformation law if the elementary systems interact. Provided that the interaction weakens when the systems are far from one another, one expects that for each mass M and spin ℓ of the combined non-interacting system there will be a corresponding state of the interacting system. There may be scattering in this state in the interacting system but the state will still be a possible state of the system with interaction. On the other hand, the interaction may produce new bound states whose mass would ordinarily be expected to be less than $m_1 + m_2$ but which, in principle, could also be greater. These would be additional elementary systems produced by the interaction.

Similar arguments are valid for three or more particles. The only new feature which appears is that, when the combined system is analyzed into elementary systems, there will be an infinity of independent states of a given mass and integer spin for every mass above the sum of the masses of the constituents; that is, the multiplicity of the representation is infinite. The argument has a straightforward extension to the case in which the combined systems have spin.

The moral of the story up to this point is that if one knew in advance what elementary systems of discrete mass a theory predicted one could write down the transformation law of the theory under $\mathscr{P}_+^\uparrow$. In a suitable basis, it would be the same transformation law as that of a system of non-interacting particles of those same discrete masses, combined in arbitrary numbers. For example, a theory describing neutral mesons of a single kind, whose mass is m, which do not form bound states, would have the spectrum shown in Figure 1–3.

Another way of expressing the idea of the preceding paragraph is to say that the collision states of the elementary systems of a theory together with the vacuum state span its Hilbert space. This can be expressed in more detail in a slightly different way as follows.

Given some set of elementary subsystems, one can form states which describe them as initially separated, then as colliding and producing outgoing waves of product elementary subsystems. Similarly, one can, by reversing velocities, describe states with ingoing waves of

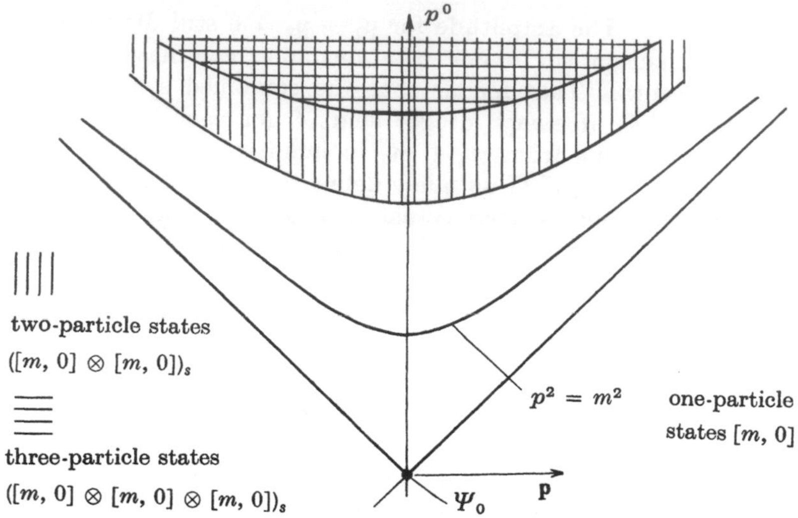

FIGURE 1–3. The spectrum of a theory of neutral scalar mesons of mass m without bound states.

products. In the limit in which the initial packets go over into plane waves these become stationary (non-normalizable!) collision states labeled by the momenta and spins of the elementary subsystems in the colliding beams. Those that satisfy the outgoing wave condition are called *in-states*; those that satisfy the ingoing wave condition *out-states*. The Hilbert spaces they span are respectively $\mathscr{H}_{\text{in}}$ and $\mathscr{H}_{\text{out}}$. The transformation properties of in-states and out-states can be deduced from those of the beams which occur in them and are just those described above. The slightly sharpened form of the above hypothesis about $U(a, \Lambda)$ is therefore $\mathscr{H}_{\text{in}} = \mathscr{H}_{\text{out}} = \mathscr{H}$. This is usually referred to as *asymptotic completeness*. For an asymptotically complete theory, the operator which takes an out-state of given momenta and spins into the in-state of the same momenta and spins is unitary if the states are normalized. This operator is called the S-operator (scattering operator) of the theory. The S-matrix is given by $(\Phi^{\text{out}}, \Psi^{\text{in}}) = (\Phi^{\text{in}}, S\Psi^{\text{in}})$.

It is worth noting that if one has once accepted the condition of asymptotic completeness, a notion of *asymptotic fields* (in- and out-fields) is already uniquely fixed whether or not one is dealing with a field theory. (A precise notion of field will not be defined until Chapter 3, so we content ourselves with a brief remark here.) We need only define creation and annihilation operators for the in-fields

$a_r^{\text{in}}(p)^*$ and $a_r^{\text{in}}(p)$, respectively. Applied to an in-state, they respectively map it onto a state with one more particle (of momentum p and spin characterized by r) and one less. Then the in-field is

$$\varphi_\alpha^{\text{in}}(x) = \int d\Omega_m(p)\left[\sum_r \Psi_\alpha(p, r)e^{-ip\cdot x}a_r(p) + \sum_r \Psi_\alpha^c(p, r)e^{+ip\cdot x}a_r(p)^*\right],$$

where Ψ is a c-number spinor quantity. [For simplicity we have considered a particle which is its own anti-particle; otherwise $\varphi_\alpha^{\text{in}}(x)$ would not be equal to its charge conjugate and $a_r(p)^*$ in this expression would have to be replaced by $b_r(p)^*$, the creation operator for the anti-particle.] One gets directly from the transformation law of the in-states

$$U(a, A)\varphi_\alpha^{\text{in}}(x)U(a, A)^{-1} = \sum_{\alpha'} S_{\alpha\alpha'}(A^{-1})\varphi_{\alpha'}^{\text{in}}(Ax + a).$$

It also follows that if the set of all in-fields is irreducible, the operators $U(a, A)$ can be written as the same functions of in-fields as for a theory of free fields. Analogous statements hold for out-fields. From the definition of the S-operator it follows that $S^{-1}\varphi^{\text{in}}(x)S = \varphi^{\text{out}}(x)$.

RELATION TO THE GENERAL ANALYSIS OF RELATIVISTIC INVARIANCE†

The preceding discussion makes it reasonable that any relativistically invariant theory in which the states are spanned by the collision states of the elementary systems of the theory has, in a suitable basis, an essentially uniquely determined relativistic transformation law. This transformation law is identical to that of a theory of non-interacting elementary subsystems of the same masses and spins. Any relativistic theory of particles which does not have this transformation law will, in our opinion, require a novel physical interpretation. (As usual, in making this statement we are ignoring the special difficulties associated with zero mass particles.)

It is of some interest to compare the results of the preceding inductive discussion with the results of a general analysis of relativistic invariance based on only the most general requirements of symmetry under $\mathscr{P}_+^\uparrow$.

Theorem 1–1 tells us that for a system for which $\{a, \Lambda\}$ is a symmetry, there is a unitary or anti-unitary operator $U(a, \Lambda)$ which is unique up to a factor on each coherent subspace. In the systematic study we must start from the ray correspondence and analyze the physical significance of the arbitrary phases in $U(a, \Lambda)$. Briefly, this analysis goes as follows.

† Further details are to be found in Refs. 5, 6, and 7.

Suppose the operations $\{a, \Lambda\}$ of $\mathscr{P}_+^\uparrow$ define symmetries. For simplicity we shall assume that the transition probabilities are continuous in the parameters of the group. This means that if $\mathbf{\Phi}$ and $\mathbf{\Psi}$ are physically realizable states and if $\mathbf{\Psi}_{a\Lambda}$ is the state resulting from the operation $\{a, \Lambda\}$ applied to $\mathbf{\Psi}$, then

$$|(\mathbf{\Phi}, \mathbf{\Psi}_{a\Lambda})|^2 \tag{1–60}$$

is continuous in the parameters (a, Λ). It follows from this hypothesis and the structure of $\mathscr{P}_+^\uparrow$ that the operators $U(a, \Lambda)$ carry coherent subspaces into themselves and are unitary rather than anti-unitary. To see this, note that the identity transformation $U(0,1)$ corresponds to leaving $\mathbf{\Psi}$ alone, and therefore cannot possibly permute the coherent subspaces. The same can be said for all $\{a, \Lambda\}$ sufficiently close to the identity from the continuity of (1–60). Since $\mathscr{P}_+^\uparrow$ is connected, this argument can be repeated in a family of neighborhoods leading to an arbitrary element of $\mathscr{P}_+^\uparrow$. That $U(a, \Lambda)$ is unitary rather than anti-unitary can be seen from the remark that the square of an anti-unitary operator is unitary, and every translation or restricted Lorentz transformation can obviously be written as a square.

Thus, within a coherent subspace, invariance under $\mathscr{P}_+^\uparrow$ implies the existence of a family of unitary operators U depending on parameters (a, Λ). The uniqueness of $U(a, \Lambda)$ up to a factor leads to the relation

$$U(a_1, \Lambda_1)U(a_2, \Lambda_2) = \omega(a_1, \Lambda_1; a_2, \Lambda_2)U(a_1 + \Lambda a_2, \Lambda_1\Lambda_2),$$

where $|\omega| = 1$. There is no change in the physical content of this transformation law if $U(a,\Lambda)$ is replaced by $e^{i\alpha(a,\Lambda)}U(a,\Lambda)$. This replaces $\omega(a_1\Lambda_1; a_2, \Lambda_2)$ by $\exp i[\alpha(a_1, \Lambda_1) + \alpha(a_2, \Lambda_2) - \alpha(a_1 + \Lambda_1 a_2, \Lambda_1\Lambda_2)]\omega$. It is natural to ask whether by a suitable choice of α one can get rid of ω altogether. The well-known answer is: it is possible to arrange $\omega = \pm 1$ and, by replacing $\mathscr{P}_+^\uparrow$ by the inhomogeneous group of $SL(2,C)$, one can arrange $\omega = +1$, and that $\{a, A\} \to U(a, A)$ is a continuous unitary representation. We quote without proof the following theorem of Wigner.

Theorem 1–2

> Every continuous unitary representation up to a factor of $\mathscr{P}_+^\uparrow$ can be brought by a suitable choice of phase factor into the form of a continuous representation $\{a,A\} \to U(a,A)$ of the inhomogeneous $SL(2,C)$.

Our next task is to summarize the results of the mathematical analysis of these continuous unitary representations of the inhomogeneous $SL(2,C)$. Every continuous unitary representation $\{a, A\} \rightarrow U(a, A)$ is unitary equivalent to one which is decomposed into irreducible representations. Two representations are unitary equivalent if the measures that tell which irreducible representations occur in the decomposition yield zero for the same subsets of irreducible representations, and if the multiplicity functions which tell how many times a given irreducible representation occurs are equal. The irreducible representations are labeled by several parameters, the first of which labels the momenta which occur in the states of the representation. [The notion of energy-momentum can be defined purely group theoretically because every continuous unitary representation of the translation group is of the form $U(a, 1) = \exp iP^\mu a_\mu$, where the P^μ are commuting self-adjoint operators.] There are six cases: $p^2 = m^2 > 0$, $p^0 \lessgtr 0$; $p^2 = 0$, $p^0 \lessgtr 0$; $p^2 = -m^2 < 0$; and $p = 0$. For the first two cases the representations are just those given above and labeled $[m,s]$, $s = 0, \frac{1}{2}, 1, \frac{3}{2}, \ldots$, and analogous ones with negative energy states. For the second two cases, we have representations $[0,s]$ with $s = 0, \pm\frac{1}{2}, \pm 1, \ldots$ where the $\pm$ sign describes the helicity of the particle and $|s|$ its spin. For example, $[0, \pm\frac{1}{2}]$ gives the transformation law of neutrinos or anti-neutrinos of the respective helicities while $[0, \pm 1]$ is the transformation law for a right (left) circularly polarized photon. There are analogous representations of zero mass and negative energy. There are other representations of zero mass but they describe systems whose spin is infinite. The remaining representations have either imaginary mass, i.e., $p^2 < 0$, or null momentum. The only one of these latter transformation laws which has an obvious physical interpretation is the identity representation which represents all Lorentz transformations by 1 on a one-dimensional Hilbert space. It describes the transformation law of the vacuum state.

Now we are in a position to see what is the nature of the restriction on the transformation law implied by the assumptions of the preceding section. It comes in three parts.

1. No states of negative energy exist. Therefore, no irreducibles of $m \geqslant 0$ and negative energy, and no irreducibles of imaginary mass occur in the transformation law.

2. There is a unique state, the vacuum, with $p = 0$. Thus, the only irreducible representation with $p = 0$ which occurs is the identity representation and it occurs with multiplicity one.

3. There is a set of irreducible representations of definite mass $[m_i, s_i]$ $i = 1, 2, \ldots$ such that all multiplicities of other irreducible representations *are those which occur in a theory of non-interacting particles of those masses and spins.*

It is implicit in the third assumption that we exclude zero mass infinite spin representations as not occurring in Nature.

In most of this book we shall use only (*1*) and (*2*); (*3*) is a vital part of all deeper investigations into collision theory.

BIBLIOGRAPHY

The present chapter is about what is sometimes, and with justice, called Wignerism: the theory of symmetry in quantum mechanics. The original sources are:

1. E. P. Wigner, *Gruppentheorie und ihre Anwendung auf die Quantenmechanik der Atomspektren*, Friedr. Vieweg Braunschweig, 1931; English translation, *Group Theory and Its Application to the Quantum Mechanics of Atomic Spectra*, Academic, New York, 1959. The general method for analyzing symmetry in quantum mechanics is given here and worked out for the case of the rotation group.
2. E. P. Wigner, "Über die Operation der Zeitumkehr in der Quantenmechanik," *Gött. Nach.*, 546–559 (1931). Here an anti-unitary symmetry appeared for the first time.
3. E. P. Wigner, "Unitary Representations of the Inhomogeneous Lorentz Group," *Ann. Math.*, **40**, 149 (1939).
4. G. C. Wick, E. P. Wigner, and A. S. Wightman, "Intrinsic Parity of Elementary Particles," *Phys. Rev.*, **88**, 101 (1952). The notion of superselection rule first appears here.

The application to relativistic quantum mechanics is summarized in a number of places:

5. A. S. Wightman, "Quelques problèmes mathématiques de la théorie quantique relativiste," pp. 6–11 in *Les problèmes mathématiques de la théorie quantique des champs*, Centre National de la Recherche Scientifique, 1959.
6. A. Barut and A. S. Wightman, "Relativistic Invariance and Quantum Mechanics," *Nuovo Cimento Suppl.*, **14**, 81–94 (1959).
7. A. S. Wightman, "L'Invariance dans la mécanique quantique relativiste," pp. 161–226 in *Dispersion Relations and Elementary Particles*, Wiley, New York, 1960.
8. S. Schweber, *An Introduction to Relativistic Quantum Field Theory*, Part One, Harper and Row, New York, 1961.
9. B. L. van der Waerden, *Die gruppentheoretische Methode in der Quantenmechanik*, Springer, Berlin, 1932.
10. G. Ya. Liubarski, *The Application of Group Theory in Physics*, translated by S. Dedijer, Pergamon, New York, 1960.

References 5, 6, 7, and 8 contain a fairly comprehensive set of references. The earnest student is referred to Ref. 7 for a set of exercises on the inhomogeneous Lorentz group which will supplement the rather sketchy treatment given here.

CHAPTER 2

SOME MATHEMATICAL TOOLS

In the thirties, under the demoralizing influence of quantum-theoretic perturbation theory, the mathematics required of a theoretical physicist was reduced to a rudimentary knowledge of the Latin and Greek alphabets.

R. Jost

The two mathematical notions in terms of which everything in the present chapter will be expressed are distribution and holomorphic function. These are discussed in the first four sections of the chapter. The last section is devoted to a few remarks on Hilbert space.

2–1. DEFINITION OF DISTRIBUTION

Distribution is a generalization of the notion of *function*, which makes it possible to make precise various formal mathematical manipulations common among physicists. The Dirac δ-function and its derivatives are examples of distributions; they are defined by the equations

$$\begin{aligned} \int f(x)\delta(x)\,dx &= f(0) \\ \int f(x)\delta'(x)\,dx &= -f'(0) \\ &\vdots \\ \int f(x)\delta^{(n)}(x)\,dx &= (-1)^n \left.\frac{d^n f}{dx^n}\right|_{x=0}, \end{aligned} \tag{2–1}$$

where $f(x)$ is some suitably smooth function on the real line. It is clear that $\delta(x)$ is not a function [it would be $= 0$ for $x \neq 0$, which implies (for a function δ) $\int f(x)\delta(x)\,dx = 0$]. Rather, it is a rule associating a number $[f(0)]$ with each suitably smooth function f. How smooth is suitably smooth? To give the formula (2–1) meaning for all n, $f(x)$ has to have derivatives of all orders at $x = 0$. In general, we expect a different notion of distribution for each class of f's for which the distribution is supposed to be defined. An f for which the distribution is defined is called a *test function.*

The essential feature of formulas such as those of (2–1) is that they define a *linear functional*; that is, for each test function, f, there is given a complex number $T(f)$ and

$$T(\alpha f) = \alpha T(f), \qquad T(f_1 + f_2) = T(f_1) + T(f_2). \tag{2–2}$$

[In this notation (2–1) is written $\delta(f) = f(0)$, etc.] Furthermore, $T(f)$ is continuous in the sense that if f_n approaches f, $T(f_n)$ approaches $T(f)$. Just as there are several possible candidates for the sets of test functions on which a distribution may be defined, there may be several possible notions of convergence $f_n \to f$. If the set of test functions is regarded as equipped with a particular notion of convergence, then a *distribution is defined as a continuous linear functional on the test functions.*

In what sense does distribution constitute a generalization of the notion of function? If T is a function defined on the same space as the test functions, it may happen that the product $T(x)f(x)$ defines an integrable function for every test function f, and that $T(f)$, which is defined as a linear functional by

$$T(f) = \int T(x)f(x)\, dx, \tag{2–3}$$

is continuous in f. In that case the function T defines a distribution. Conversely, we say a distribution T *is* a function if there exists a function such that (2–3) holds. Clearly, in this correspondence between distributions and functions, one does not distinguish between two functions which differ on a set of measure zero because each will give rise to the same linear functional (2–3).

Much of the ordinary terminology of function theory can be transferred immediately to distributions. For example, there is the notion of the *support* of a function, T, which will be denoted† supp T. It is the closed set obtained by taking the complement of the largest open set in which T vanishes. Now, to say that T vanishes in an open set $\mathcal{O}$ is nearly equivalent for a function to saying that $T(f) = 0$ for all test functions whose supports are in $\mathcal{O}$, provided the test functions are numerous enough to include sufficiently many positive functions whose supports are in $\mathcal{O}$. (We say nearly equivalent because there may be exceptional sets of measure zero.) Thus we adopt as the definition of support for a distribution T: supp T is the complement of the largest open set on which T vanishes. T vanishes on an open set if it vanishes for all test functions whose supports are in the open set. This definition gives meaning to the notion of vanishing in a neighborhood of a point but not *at* a point. For all our applications the former will be adequate.

† Supp T should not be confused with sup T, which is a shorthand for the least upper bound, or supremum, of T.

With the exception of two important results, we shall deal in this book with the special case of so-called *tempered distributions*. The set of test functions in this case is denoted $\mathscr{S}$ or, when it is advisable to indicate on what variables the test functions depend, $\mathscr{S}(\mathbf{R}^n)$ or $\mathscr{S}_{x_1,\ldots x_n}$. Here $\mathbf{R}^n$ stands for the real vector space of n dimensions equipped with the usual euclidean distance

$$|x - y| = \left[\sum_{j=1}^{n} (x_j - y_j)^2\right]^{1/2}.$$

$\mathscr{S}$ consists of all complex-valued, infinitely differentiable functions f, which, together with their derivatives, approach zero at infinity faster than any power of the euclidean distance. This last requirement can be made explicit if we introduce the following notation. Let k stand for the sequence of integers $k_1, \ldots k_n$ and write x^k for the product of powers given by

$$x^k = x_1^{k_1} \cdots x_n^{k_n}. \tag{2–4}$$

Put D^k for the differentiation operator given by

$$D^k = \frac{\partial^{|k|}}{(\partial x_1)^{k_1} \cdots (\partial x_n)^{k_n}}, \tag{2–5}$$

where $|k| = k_1 + k_2 + \cdots k_n$. Now define

$$\|f\|_{r,s} = \sum_{\substack{k \\ |k| \leqslant r}} \sum_{\substack{\ell \\ |\ell| \leqslant s}} \sup_x |x^k D^\ell f(x)|. \tag{2–6}$$

For each pair of non-negative integers r and s, and each infinitely differentiable function f, this formula yields an $\|f\|_{r,s}$ which is either a non-negative real number or $+\infty$. If $\|f\|_{r,s} = 0$ for some pair r, s, then $f = 0$ because that implies $\sup_x |f(x)| = 0$. Furthermore, $\|\alpha f\|_{r,s} = |\alpha| \, \|f\|_{r,s}$ and $\|f + g\|_{r,s} \leqslant \|f\|_{r,s} + \|g\|_{r,s}$, so $\| \quad \|_{r,s}$ is a norm. *The class $\mathscr{S}$ is the set of all infinitely differentiable functions, f, such that*

$$\|f\|_{r,s} < \infty$$

for all integers r, s. It is pretty evident that this definition is equivalent to the preceding one.

To define convergence in $\mathscr{S}$, one again uses the $\| \quad \|_{r,s}$. A sequence of f_n in $\mathscr{S}$ converges to f in $\mathscr{S}$ if, for each r and s,

$$\lim_{n\to\infty} \|f_n - f\|_{r,s} = 0. \tag{2–7}$$

This machinery gives a neat statement of the definition of tempered distribution. A tempered distribution, T, is a linear functional defined on $\mathscr{S}$ with the property: if

$$\lim_{n\to\infty} \|f_n - f\|_{r,s} = 0 \qquad \text{for all } r, s$$

then

$$\lim_{n\to\infty} |T(f_n) - T(f)| = 0. \tag{2–8}$$

Just as in the case of a function, F, of a real variable, the definition of continuity can be stated in alternative forms. For a function,

$$x_n \to x \qquad \text{implies} \qquad F(x_n) \to F(x),$$

which is equivalent to: for any $\varepsilon > 0$ there exists a $\delta > 0$ such that

$$|y - x| < \delta \qquad \text{implies} \qquad |F(y) - F(x)| < \varepsilon.$$

For a distribution the analogue of the second alternative is: the tempered distribution T is continuous at f if, for any $\varepsilon > 0$, there exist integers r, s, and a $\delta > 0$, such that

$$\|g - f\|_{r,s} < \delta \qquad \text{implies} \qquad |T(g) - T(f)| < \varepsilon. \tag{2–9}$$

We shall use whichever of the equivalent formulations (2–8) or (2–9) is convenient.

One way to guarantee that (2–8) or (2–9) holds is to require that for some pair of non-negative integers r, s there exists a constant C such that

$$|T(f)| \leqslant C\|f\|_{r,s}. \tag{2–10}$$

[Clearly then,

$$|T(f_n) - T(f)| = |T(f_n - f)| \leqslant C\|f_n - f\|_{r,s}.]$$

Actually, this condition (2–10) is not only sufficient for the continuity of T but also necessary, as is easy to prove using the linearity of T. [We shall not prove this but shall use it; it is Theorem (5–1), p. 25, in Gårding–Lions.[3] A completely analogous assertion for operators in Hilbert space is proved in Section 2–6.] Incidentally, (2–10) shows that a linear functional is continuous at all points f in $\mathscr{S}$ if it is continuous at $f = 0$.

There is an important class of linear functionals which satisfy the continuity condition (2–10) in an obvious way:

$$T(f) = \sum_{0\leqslant|k|\leqslant s} \int F_k(x_1,\ldots x_n) D^k f(x_1,\ldots x_n)\, dx_1\ldots dx_n, \tag{2–11}$$

where the F_k are continuous functions bounded as follows:

$$|F_k(x)| \leqslant C_k(1 + |x|^j) \tag{2–12}$$

for some C_k and j depending on k. Clearly, it then follows that

$$|T(f)| \leqslant C\|f\|_{r,s}$$

for some C and r, s, so T is a tempered distribution. It is usually written symbolically in the form obtained by integrating by parts:

$$T(x) = \sum_{0 \leqslant |k| \leqslant s} (-1)^{|k|} D^k F_k(x). \tag{2–13}$$

In general, this formula cannot be taken literally, since the F_k need not be differentiable nor the $D^k F_k$ integrable.

It can be shown that *every* tempered distribution can be written in the form (2–11). We shall not prove this result here. (It is an exercise on p. 28 of Gårding–Lions.) We merely note that the practical-minded reader can take (2–11) as the *definition* of what is meant by a tempered distribution, and will not go wrong in any of what follows.

It is clear that the tempered distributions are very restricted in two ways, which can be described roughly as follows: they grow at worst like polynomials when $x \to \infty$, and they are at worst of finite order. The first means essentially that the inequalities (2–12) hold and only a finite number of terms occur in (2–11), whereas the second means that D^k only up to some maximum $|k|$ occur in (2–11). Of course, precise definitions of polynomial growth and order can be given, but we shall not need them.

The only distributions appearing in this book which are not tempered will be elements of $\mathscr{D}'$, that is, continuous linear functionals on the space $\mathscr{D}$ of all infinitely differentiable functions of compact support.† The notion of convergence in $\mathscr{D}$ is this: $f_n \to f$ in $\mathscr{D}$, if the supports of the f_n lie in some fixed compact set, K, and $f_n \to f$ uniformly in K, and the derivatives of f_n approach those of f uniformly in K.

Clearly, every element of $\mathscr{D}$ is an element of $\mathscr{S}$, so every element of $\mathscr{S}'$ is defined on the elements of $\mathscr{D}$. Furthermore, a sequence of elements of $\mathscr{D}$ which converges in $\mathscr{D}$ to an element of $\mathscr{D}$ certainly converges in $\mathscr{S}$ to the same element. Therefore $\mathscr{S}' \subset \mathscr{D}'$, or, in ordinary language, tempered distributions are distributions. However, there are many elements of $\mathscr{D}'$ which are not in $\mathscr{S}'$. For one example, take ordinary continuous functions which grow exponentially. For another, consider

$$T(x) = \sum_{n=0}^{\infty} \delta^{(n)}(x - n).$$

† Recall that a set, S, is compact if given a collection $\{\mathcal{O}_i\}$ of open sets such that $x \in S$ implies $x \in \mathcal{O}_i$ for some i; then there is a finite subset of the collection with the same property. More briefly, S is compact if every open covering has a finite subcovering. For sets in euclidean space, a set is compact if and only if it is closed and bounded.

When applied to any test function in $\mathscr{D}$, the infinite sum becomes finite. However, there is no bound on the order of the derivative which appears if the support of the test function moves toward infinity; elements of $\mathscr{D}'$ need not be of finite order.

In both $\mathscr{S}'$ and $\mathscr{D}'$, we can define what is meant by the *convergence of a sequence*, T_n, *of distributions to a distribution* T: $T_n \to T$ if, for every test function f, $T_n(f) \to T(f)$ as complex numbers. Of course, the notions of convergence are not the same because the f's are different. A typical example of convergence of distributions occurring in practice is this. Suppose for each n, τ_n is a continuous mapping of the test functions into the test functions, and furthermore, $\tau_n(f) \to \tau(f)$ for each test function f, where $\tau(f)$ is another continuous mapping of the test functions into themselves. Then, for a fixed distribution V, we define $T_n(f) = V[\tau_n(f)]$ and $T(f) = V[\tau(f)]$. Then $T_n \to T$.

The notion of convergence for a sequence of distributions can be treated more systematically if $\mathscr{S}$, $\mathscr{S}'$, $\mathscr{D}$, and $\mathscr{D}'$ are characterized as topological vector spaces. (In its essentials, we have already done that for $\mathscr{S}$ by introducing the family of norms $\| \quad \|_{r,s}$.) We refer the reader to Refs. 1, 2, and 3 for a full development of this point of view. Here, we shall only make three remarks along these lines which are essential for our later exposition. The first is that $\mathscr{S}$ and $\mathscr{D}$ are *separable* as topological spaces. The required dense sequence of vectors may be constructed as follows. Take all polynomials in the variables $x_1, \ldots x_n$ with coefficients whose real and imaginary parts are rational numbers, and multiply them by a sequence of infinitely differentiable functions which are 1 inside a sphere of radius n, and 0 outside a sphere of radius $n + 1$. The resulting set is countable and Weierstrass's theorem, saying that one can approximate a continuous function on a compact set by polynomials, makes plausible that it has every vector in $\mathscr{S}$ and $\mathscr{D}$ as a limit point. This will not be shown here. (See Ref. 24, p. 373, for a proof.)

The second remark is that $\mathscr{D}'$ and $\mathscr{S}'$ are complete; i.e., if T_k is a sequence of distributions such that $T_k(f)$ is a Cauchy sequence of real numbers for each test function f, then there exists a distribution T such that $T_k \to T$.

The third remark is that a central role in the description of topological vector spaces is played by the notion of *bounded set*. A set S in a topological vector space is bounded if, for each neighborhood N of 0 in the vector space, there exists a real number $\lambda > 0$ such that $\lambda S \subset N$. The bounded sets of $\mathscr{S}$, $\mathscr{D}$, $\mathscr{S}'$, and $\mathscr{D}'$ can be characterized as follows. A set $S \subset \mathscr{S}$ is bounded if and only if all $\| \quad \|_{r,s}$ are bounded functions on S. A set $S \subset \mathscr{D}$ is bounded if and only if there exists a fixed compact set K such that $f \in S$ implies $\operatorname{supp} f \subset K$, and for each non-negative integer k there is a real number, M_k, such that $\sup_{x \in K} |D^k f(x)| \leqslant M_k$ for $f \in S$. A set $S \subset \mathscr{S}'$ is bounded if, for each $f \in \mathscr{S}$, $T(f)$ is bounded as T varies over S. The same statement holds with $\mathscr{S}'$ replaced by $\mathscr{D}'$ and $\mathscr{S}$ by $\mathscr{D}$. The only result involving bounded sets which we shall need later is this: a convergent sequence of distributions $\in \mathscr{D}'$ converges uniformly on bounded sets of $\mathscr{D}$. We shall use this in Section 2–5 in the following way. We shall have a sequence, T_n, of distributions which converge for each test function of the form $f_a(x) = f(x - a)$, where a runs over some region $|a| \leqslant \rho$. Then, because the f_a form a bounded set in $\mathscr{D}$, we shall have that the sequence converges uniformly in a. (See Ref. 1, p. 74 for a proof.)

One last generality about the definition of distribution. We have carried out the preceding discussion for distributions defined on $\mathscr{D}$ and $\mathscr{S}$, where the domain of the test functions is all of $\mathbf{R}^n$. The whole discussion can also be carried out for test functions defined only on an open subset $\mathscr{O}$ of $\mathbf{R}^n$. We shall not spell out all the details here but only note that the elements of $\mathscr{D}(\mathscr{O})$ are infinitely differentiable and have supports which are compact subsets of $\mathbf{R}^n$ contained in $\mathscr{O}$. The continuous linear functionals $\mathscr{D}(\mathscr{O})'$ on $\mathscr{D}(\mathscr{O})$ will be used in Section 2–5.

MISCELLANEOUS PROPERTIES OF DISTRIBUTIONS

One of the main motivations for the introduction of the notion of distribution was the desire to have a class of objects for which differentiation is always permissible. The derivative, $\partial T/\partial x_j$, of a distribution, T, is defined by

$$\frac{\partial T}{\partial x_j}(f) = -T\left(\frac{\partial f}{\partial x_j}\right). \tag{2–14}$$

That (2–14) defines a distribution is an immediate consequence of the definitions given above; if a sequence f_n converges to f in $\mathscr{S}$ (in $\mathscr{D}$), then $\partial f_n/\partial x_j$ converges to $\partial f/\partial x_j$ in $\mathscr{S}$ (in $\mathscr{D}$). It immediately follows that

$$D^p T(f) = (-1)^{|p|} T(D^p f). \tag{2–15}$$

One can equivalently define the derivative of T as the limit in $\mathscr{S}'$ (or $\mathscr{D}'$),

$$\lim_{a_j \to 0} (a_j)^{-1}(T_{a_j} - T), \tag{2–16}$$

where T_{a_j} is T translated in its jth coordinate, a notion which will be defined shortly. According to the definition of the notion of convergence in $\mathscr{S}'$ (or $\mathscr{D}'$), (2–16) means that for each $f \in \mathscr{S}$ (or $\in \mathscr{D}$)

$$\frac{\partial T}{\partial x_j}(f) = \lim_{a_j \to 0} (a_j)^{-1}(T_{a_j} - T)(f). \tag{2–17}$$

But this is

$$\lim_{a_j \to 0} (a_j)^{-1}[T_{a_j}(f) - T(f)] = \lim_{a_j \to 0} T[a_j^{-1}(f_{-a_j} - f)]$$
$$= -T\left(\frac{\partial f}{\partial x_j}\right),$$

where

$$f_{-a_j}(x) = f[x - (0, \ldots a_j, \ldots 0)],$$

because $a_j^{-1}[f_{-a_j}(x) - f(x)]$ converges to $-(\partial f/\partial x_j)(x)$ in $\mathscr{S}$ (or $\mathscr{D}$).

The operation of translation of the jth argument used in this alternative definition of derivative is a special case of an arbitrary nonsingular inhomogeneous linear transformation of $\mathbf{R}^n$:

$$x \to Lx + a,$$

which we denote $\{a, L\}$. We write for $f \in \mathscr{S}$ (or $\in \mathscr{D}$)

$$(\{a, L\}f)(x) = f(L^{-1}(x - a))$$

and define for distributions $\in \mathscr{S}'$ (or $\in \mathscr{D}'$)

$$T_{\{a,L\}}(f) = |\det L|^{-1} T(\{a, L\}f).$$

It is elementary to verify that this defines a distribution in $\mathscr{S}'$ (or $\mathscr{D}'$), respectively. The definition has been so arranged that if T is a function,

$$T_{\{a,L\}}(x) = T(Lx + a).$$

Notice that for $f \in \mathscr{S}$ (or $\in \mathscr{D}$)

$$\{a, L\}(\{b, M\}f) = \{a + Lb, LM\}f$$

and for $T \in \mathscr{S}'$ (or $\in \mathscr{D}'$)

$$(T_{\{a,L\}})_{\{b,M\}} = T_{\{a+Lb,LM\}}.$$

The invariance of a distribution under $\{a, L\}$ is defined in the expected way

$$T_{\{a,L\}} = T.$$

A particular case of this kind of transformation occurs in the treatment of expectation values in Section 3–3. There one has tempered distributions on $\mathbf{R}^{4n}$ and describes a point of $\mathbf{R}^{4n}$ by a set of n four-vectors $x_1, \ldots x_n$. The vacuum expectation values are invariant under the linear transformation

$$x_1, \ldots x_n \rightarrow x_1 + a, \ldots x_n + a, \tag{2–18}$$

where a is any real four-vector. We are led to the problem of describing all tempered distributions with this invariance property. For the special case for which the distribution is a function, the obvious answer is that T depends only on the difference variables $x_1 - x_2, \ldots x_{n-1} - x_n$. This also is true for distributions but requires some argument which we now give. First, we make the non-singular linear transformation

$$(x_1, \ldots x_n) \rightarrow (X, \xi_1, \xi_2, \ldots \xi_{n-1})$$

where

$$X = n^{-1}[x_1 + \cdots + x_n]; \quad \xi_1 = x_1 - x_2, \ldots \xi_{n-1} = x_{n-1} - x_n, \tag{2–19}$$

which carries the given distribution T into a distribution T_1. The invariance of T under (2–18) is equivalent to the invariance of T_1 under

$$X, \xi_1, \ldots \xi_{n-1} \rightarrow X + a, \xi_1, \ldots \xi_{n-1}. \tag{2–20}$$

This invariance is, in turn, equivalent to the property

$$\frac{\partial T_1}{\partial X_\mu} = 0 \qquad \mu = 0, 1, 2, 3. \tag{2–21}$$

That invariance under (2–20) implies (2–21) is evident from the definition (2–17) of the derivative. Conversely, if the derivatives with respect to X_μ vanish, we can proceed as follows. For a fixed $f_0 \in \mathscr{S}(\mathbf{R}^1)$ (or $\in \mathscr{D}(\mathbf{R}^1)$) with the property $\int f_0(x)\, dx = 1$, define for each $f \in \mathscr{S}(\mathbf{R}^{4n})$ (or $\mathscr{D}(\mathbf{R}^{4n})$)

$$\begin{aligned}\chi(X, \xi_1, \ldots \xi_{n-1}) &= f(X, \xi_1, \ldots \xi_{n-1}) \\ &\quad - f_0(X_0) \int dy\, f(y, X_1, X_2, X_3, \xi_1, \ldots \xi_{n-1}).\end{aligned} \tag{2–22}$$

χ is also in $\mathscr{S}(\mathbf{R}^{4n})$ (or $\mathscr{D}(\mathbf{R}^{4n})$) and satisfies

$$\int \chi(y, X_1, X_2, X_3, \xi_1, \ldots \xi_{n-1})\, dy = 0. \tag{2–23}$$

Now a test function χ satisfying (2–23) can always be written in the form $\frac{\partial}{\partial X_0} \chi_1$, where

$$\chi_1(X_0, \ldots X_3, \xi_1, \ldots \xi_{n-1}) = \int_{-\infty}^{X_0} dy\, \chi(y, X_1, X_2, X_3, \xi_1, \ldots \xi_{n-1}).$$

Therefore,

$$T_1(\chi) = T_1\left(\frac{\partial}{\partial X_0}\chi_1\right) = -\frac{\partial T_1}{\partial X_0}(\chi_1) = 0 \tag{2–24}$$

and so $T_1(f)$ can be written

$$T_1(f) = T_1\left(f_0(X_0)\int dy\, f(y, X_1, X_2, X_3, \xi_1, \ldots \xi_{n-1})\right)$$

$$= T_2\left(\int dy\, f(y, X_1, X_2, X_3, \xi_1, \ldots \xi_{n-1})\right) \tag{2–25}$$

T_2 is a distribution in $X_1, X_2, X_3, \xi_1, \ldots \xi_{n-1}$. When X_1, X_2, X_3 have been treated in the same way, we have given a precise meaning to the equation $T(x_1, \ldots x_n) = T(\xi_1, \ldots \xi_{n-1})$ and have solved the problem of finding all distributions invariant under (2-18).

Another operation which is defined for functions and has a natural extension to distributions is *multiplication by a function* (see Ref. 1, Chapter 5). We propose to define for a distribution T and function g

$$(gT)(f) = T(gf), \tag{2–26}$$

where $(gf)(x) = g(x)f(x)$. In order that (2–26) should define a distribution, we want g to be such that gf is a test function depending continuously on f. For this it suffices, if $f \in \mathscr{D}$, that g be infinitely differentiable, since then $D^p(gf)$ is a linear combination of products of derivatives of g and f, which approaches zero in $\mathscr{D}$ when f does. If $f \in \mathscr{S}$, an arbitrary infinitely differentiable function g will not do because it might grow too fast. However, if g and all its derivatives are bounded by polynomials; that is, if for each p

$$|D^p g(x)| \leqslant |P_p(x)| \qquad \text{for all } x,$$

then it is easy to see that gf_n will converge to gf in $\mathscr{S}$ if f_n converges to f. Under these respective restrictions on g, the mappings $f \to gf$ of $\mathscr{D}$ into $\mathscr{D}$ or $\mathscr{S}$ into $\mathscr{S}$ are continuous, and gT is in $\mathscr{D}'$ ($\mathscr{S}'$) if T is in $\mathscr{D}'$ ($\mathscr{S}'$).

Those tempered distributions T which happen to be of the form gT_1, where $g \in \mathscr{S}$ and T_1 is also a tempered distribution, are of a special breed, *tempered distributions of fast decrease*. We shall meet them again in Section 2–2.

Quite another kind of operation is that of multiplication of a function, S, of one variable by another T depending on another variable: $S(x)T(y)$. This defines the *tensor product*:

$$(S \otimes T)(x, y) = S(x)T(y).$$

(See Ref. 1, Chapter 4.) It is trivial to extend this definition to the case in which S and T are distributions if one only defines the tensor product as a functional of two variables

$$(S \otimes T)(f, g) = S(f)T(g) \tag{2–27}$$

with f and g test functions. It requires more argument to show that (2–27) determines a distribution in both variables together, giving meaning to the expression

$$(S \otimes T)(h) = \iint h(x, y)\, dx\, dy\, S(x)T(y).$$

That such a distribution is unique, if it exists, follows from the fact that sums of the form

$$\textstyle\sum_i f_i(x)g_i(y)$$

are dense in $\mathscr{D}_{x,y}$ and $\mathscr{S}_{x,y}$. (To prove this, one can use the functions described in connection with the discussion of the separability of $\mathscr{D}$ and $\mathscr{S}$.) The existence of $S \otimes T$ is established by noting that one can evaluate successively

$$S_x(T_y(h(x, y))), \tag{2–28}$$

where the x and y indicate the association of variables in test function and distribution, to obtain a distribution which satisfies (2–27).

The last operation we shall need is *convolution*. It is defined for two elements f and g of $\mathscr{S}$ (or $\mathscr{D}$) by

$$(f * g)(x) = \int f(x - \xi)g(\xi)\, d\xi. \tag{2–29}$$

It is not difficult to see that $f * g$ is again in $\mathscr{S}$ (or $\mathscr{D}$). Furthermore, it is not difficult to verify directly from the definitions that $f * g$ is a continuous in f and g. The $*$ operation is commutative,

$$(f * g)(x) = \int f(y)g(x - y)\, dy = (g * f)(x). \tag{2–30}$$

The operation can be extended to the case in which one of the factors is a distribution, T, and the other is a test function, f, by the equation

$$(f * T)(h) = T(\check{f} * h), \tag{2–31}$$

where $\check{f}(x) = f(-x)$. This agrees with (2–29) if T happens to be a function in $\mathscr{S}$ or $\mathscr{D}$ because

$$\begin{aligned}(f * g)(h) &= \int dx[\int f(x - \xi)g(\xi)\, d\xi]h(x)\\ &= \int d\xi[\int dx\, f(x - \xi)h(x)]g(\xi)\\ &= \int d\xi[\int dx\, \check{f}(\xi - x)h(x)]g(\xi) = g(\check{f} * h).\end{aligned}$$

It should be noted that still another way of writing (2–31) is

$$(f * T)(h) = (f \otimes T)(h_1)$$

where $h_1(x, y) = h(x + y)$. That $f \otimes T$ is defined on h_1 is not immediately obvious, since h_1 belongs to neither $\mathscr{D}$ nor $\mathscr{S}$ in x and y together. For $f \in \mathscr{D}$, the expression can, nevertheless, be given a meaning as follows. Let $\chi \in \mathscr{D}$ be a function of x such that $\chi(x) = 1$ if $x \in \operatorname{supp} f$. Then $(f * T)(h) = (f\chi * T)(h) = (f \otimes T)(h_2)$, where $h_2(x, y) = \chi(x)h(x + y) \in \mathscr{D}_{x,y}$ and depends continuously on h. An analogous but more delicate argument works for $\mathscr{S}$. (See Ref. 1, Vol. II, p. 102.) That the right-hand side of (2–31) defines a continuous linear functional of h follows from the fact that $f * h$ is a continuous linear functional of h with values in $\mathscr{S}$ or $\mathscr{D}$, depending on which is under consideration.

Actually, for $T \in \mathscr{S}'$ (or $\in \mathscr{D}'$) and $f \in \mathscr{S}$ (or $\in \mathscr{D}$), $f * T$ is an infinitely differentiable function. The operation of passing from T to $f * T$ is a very important one in practice; it is known as *regularization* (by f). To see that $f * T$ is an infinitely differentiable function notice that $T(\hat{f}_{-x})$ is infinitely differentiable in x: $[\hat{f}_{-x}(\xi) = \hat{f}(\xi - x) = f(x - \xi)]$, and it and $f * T$ evaluated for the test function h are given by

$$(f * T)(h) = T(\hat{f} * h) = (T \otimes h)(\hat{f}_{-x})$$
$$= \int T(\hat{f}_{-x})\, dx\, h(x).$$

The last equality is legitimate because by construction (2–28) of the tensor product it can be evaluated "successively."

THE SCHWARTZ NUCLEAR THEOREM

In a number of practical situations, for example, those encountered in Sections 2–5 and 3–3, one has to deal with *separately* continuous multi-linear functionals defined on $\mathscr{S}$ or $\mathscr{D}$. These are complex-valued functions T of arguments $f_1, \ldots f_k \in \mathscr{S}$ or $\mathscr{D}$, satisfying

$$T(f_1, \ldots \alpha f_j' + \beta f_j'', \ldots f_k) = \alpha T(f_1, \ldots f_j', \ldots f_k) + \beta T(f_1, \ldots f_j'', \ldots f_k) \quad (2\text{–}32)$$

for $j = 1, \ldots k$; T is a distribution in each f_j with all the other arguments held fixed. Of course, an obvious way to construct such an object is to take a distribution, G, in all the variables together and then specialize it to test functions which are products of the form $f(x_1, \ldots x_k) = f_1(x_1)f_2(x_2)\ldots f_k(x_k)$. It was a remarkable discovery of Schwartz that there are no others than these; every T is of the form $T(f_1, \ldots f_k) = G(f_1 f_2 \ldots f_k)$. G is referred to as the nucleus or kernel of the multilinear functional. This terminology is chosen in analogy with the theory of integral equations where one has integral operators of the form $Tf(x) = \int k(x, y)f(y)\, dy$, and k is referred to as the kernel of T.

The only proof of the nuclear theorem which is of a sufficiently elementary character to be mentioned here is that of Gelfand and Vilenkin, Ref. 5. However, it is only for $\mathscr{D}'$ and there does not seem to be an analogous elementary proof available for $\mathscr{S}'$. Thus, we content ourselves with a statement of the theorem.

Theorem 2–1 (*Nuclear Theorem*)

Let T be a multilinear functional of arguments $f_1, \ldots f_k \in \mathscr{S}$ $(\in \mathscr{D})$ which is continuous in each of its arguments, the others being held fixed. Then there exists a unique distribution, G, $\in \mathscr{S}'$ $(\in \mathscr{D}')$ in all the variables of $f_1, \ldots f_k$ such that

$$T(f_1, f_2, \ldots f_k) = G(f_1 f_2 \ldots f_k).$$

It is easy to see from examples that G may be quite singular in its arguments without that being reflected in the behavior of T in any one of its arguments. A case in point is $G(x, y) = -\delta'(x - y)$. Here $G(f_1 f_2) = T(f_1, f_2) = \int f_1'(x) f_2(x)\, dx = -\int f_1(x) f_2'(x)\, dx$, so $T(f_1, f_2)$ is an infinitely differentiable function of each of its arguments, the other being held fixed; namely,

$$T(x_1, f_2) = -f_2'(x_1) \qquad \text{and} \qquad T(f_1, x_2) = f_1'(x_2).$$

2–2. FOURIER TRANSFORMS

We define two linear transformations $\mathscr{F}$ and $\overline{\mathscr{F}}$ on $\mathscr{S}$ by the equations

$$(\mathscr{F} f)(p) = \frac{1}{(2\pi)^{n/2}} \int e^{-ip\cdot x} f(x)\, dx \tag{2–33}$$

$$(\overline{\mathscr{F}} f)(p) = \frac{1}{(2\pi)^{n/2}} \int e^{ip\cdot x} f(x)\, dx. \tag{2–34}$$

Here $p \cdot x$ stands for the *non-degenerate scalar product* appropriate to the problem at hand. For example, it might be the euclidean scalar product $p \cdot x = \sum_{j=1}^{n} p^j x^j$. We shall later on deal exclusively with cases where n is a multiple of 4 and $p \cdot x$ is a sum of *Minkowski scalar products*

$$p \cdot x = \sum_{k=1}^{n/4} \sum_{\mu,\nu=0}^{3} p_{k\mu} g^{\mu\nu} x_{k\nu}$$

with $g^{00} = 1 = -g^{jj}$, $j = 1, 2, 3$ and $g^{\mu\nu} = 0$, $\mu \neq \nu$. Evidently, the integrals are uniformly and absolutely convergent and the same is true for all integrals obtained by multiplying by x^k or differentiating under the integral sign. Thus we have with a few integrations by parts

$$\mathscr{F}(D^k f)(p) = (+ip)^k(\mathscr{F}f)(p), \qquad \overline{\mathscr{F}}(D^k f)(p) = (-ip)^k(\overline{\mathscr{F}}f)(p) \quad (2\text{–}35)$$

$$\mathscr{F}[(-ix)^k f](p) = D^k(\mathscr{F}f)(p), \qquad \overline{\mathscr{F}}[(+ix)^k f](p) = D^k(\overline{\mathscr{F}}f)(p). \quad (2\text{–}36)$$

From (2–35) and (2–36) we immediately get that $\mathscr{F}$ and $\overline{\mathscr{F}}$ are continuous because, for example,

$$\begin{aligned} |p^r D^s(\mathscr{F}f)(p)| &= \left| \frac{1}{(2\pi)^{n/2}} \int \frac{e^{-ip\cdot x}}{(1+|x|^2)^t} D^r[(ix)^s f(x)](1+|x|^2)^t \, dx \right| \\ &\leqslant \sup_x \left[|D^r(x^s f(x))| (1+|x|^2)^t \right] \\ &\qquad \times \frac{1}{(2\pi)^{n/2}} \int \frac{dx}{(1+|x|^2)^t}. \end{aligned} \quad (2\text{–}37)$$

For suitably large positive t, the left-hand side is clearly bounded by a constant times some norm, $\|f\|_{p,q}$, of f: $\|\mathscr{F}f\|_{r,s} \leqslant C\|f\|_{p,q}$, which is what we mean by continuity in $\mathscr{S}$.

The crucial lemma, which is essentially Fourier's inversion theorem for functions in $\mathscr{S}$, is

Lemma

The transformations $\mathscr{F}$ and $\overline{\mathscr{F}}$ are isomorphisms of $\mathscr{S}$, i.e., they are one-to-one continuous mappings of $\mathscr{S}$ onto all of $\mathscr{S}$ such that the inverse mappings are continuous. In fact, they are each other's inverses on $\mathscr{S}$.

$$\mathscr{F}\overline{\mathscr{F}} = \overline{\mathscr{F}}\mathscr{F} = 1. \quad (2\text{–}38)$$

Proof:

We examine the integral

$$\frac{1}{(2\pi)^n} \int e^{-ip\cdot x} \, dx \int dq\, e^{iq\cdot x} f(q), \quad (2\text{–}39)$$

which, if the contention of the theorem is correct, ought to be $f(p)$. Formally, all that has to be done is interchange orders of integration and use

$$\frac{1}{(2\pi)^n} \int e^{ix\cdot(p-q)} \, dx = \delta(p-q). \quad (2\text{–}40)$$

To justify this one can proceed in many ways. We shall rewrite (2–39) as

$$\lim_{\varepsilon \to 0} \frac{1}{(2\pi)^n} \int e^{-ip\cdot x}\, dx \exp(-\varepsilon|x|^2) \int dq\, e^{iq\cdot x} f(q),\ \varepsilon > 0. \tag{2–41}$$

As long as $\varepsilon > 0$, the integral exists in both variables together so we are allowed to interchange the order. Using the integral

$$\int \exp(-\varepsilon|x|^2 + ir\cdot x)\, dx = (\pi/\varepsilon)^{n/2} \exp(-|r|^2/4\varepsilon) \tag{2–42}$$

we get that (2–41) is

$$\lim_{\varepsilon \to 0} \frac{1}{(4\pi\varepsilon)^{n/2}} \int \cdots \int \exp(-|p - q|^2/4\varepsilon) f(q)\, dq \tag{2–43}$$

It is a standard matter to show that (2–43) is $f(p)$. One argues that the contribution from outside any sphere $|p - q|^2 = R^2$ goes to zero as $\varepsilon \to 0$, so the difference between (2–43) and $f(p)$ can be estimated by

$$\left| \frac{1}{(4\pi\varepsilon)^{n/2}} \int_{|p-q| \leqslant R} \exp(-|p - q|^2/4\varepsilon) [f(q) - f(p)]\, dq \right|$$
$$\leqslant \sup_{|p-q| \leqslant R} |f(q) - f(p)| \to 0 \text{ as } R \to 0. \quad ∎$$

Now we can define the operation of Fourier transformation on $\mathscr{S}'$. We do it so as to be consistent with Parseval's formula for functions in $\mathscr{S}$:

$$\frac{1}{(2\pi)^{n/2}} \int g(p)\, dp \int e^{-ip\cdot x}\, dx\, h(x)$$
$$= \frac{1}{(2\pi)^{n/2}} \int dx\, h(x) \int e^{-ip\cdot x} g(p)\, dp, \tag{2–44}$$

which is an immediate consequence of the legitimacy of the interchange of order of integration when the integrand is as well behaved as $g(p)h(x)$. Equation (2–44) may be written

$$(\mathscr{F}h)(g) = h[\mathscr{F}(g)] \tag{2–45}$$

if we regard h as an element of $\mathscr{S}'$. This suggests the definition, for an arbitrary element, T, of $\mathscr{S}'$,

$$(\mathscr{F}T)(f) = T(\mathscr{F}f). \tag{2–46}$$

$\mathscr{F}$ so defined is clearly a linear transformation of $\mathscr{S}'$ into itself. Furthermore $\mathscr{F}$ is continuous on $\mathscr{S}'$ because if $T_n \to T$ in $\mathscr{S}'$,

$$(\mathscr{F}T_n)(f) = T_n(\mathscr{F}f) \to T(\mathscr{F}f) = (\mathscr{F}T)(f),$$

so $\mathscr{F}T_n \to \mathscr{F}T$ in $\mathscr{S}'$.

Similar statements hold for $\overline{\mathscr{F}}$ defined as a linear transformation of $\mathscr{S}'$ by

$$(\overline{\mathscr{F}}T)(f) = T(\overline{\mathscr{F}}f). \tag{2–47}$$

The connection between $\mathscr{F}$ and $\overline{\mathscr{F}}$ so defined is given by

Theorem 2–2

$\mathscr{F}$ and $\overline{\mathscr{F}}$ defined on $\mathscr{S}'$ by (2–46) and (2–47), respectively, are inverse isomorphisms of $\mathscr{S}'$, i.e., they are continuous linear one-to-one mappings of $\mathscr{S}'$ onto $\mathscr{S}'$ such that

$$\mathscr{F}\overline{\mathscr{F}} = \overline{\mathscr{F}}\mathscr{F} = 1 \tag{2–48}$$

on $\mathscr{S}'$.

Proof:

We have already established that $\mathscr{F}$ and $\overline{\mathscr{F}}$ are continuous linear mappings of $\mathscr{S}'$ into $\mathscr{S}'$. That they map $\mathscr{S}'$ onto $\mathscr{S}'$ follows from the corresponding statement for $\mathscr{S}$: if $T \in \mathscr{S}'$, then the element V of $\mathscr{S}'$, defined by

$$V(f) = T(\overline{\mathscr{F}}f)$$

has the property

$$(\mathscr{F}V)(f) = T(\mathscr{F}\overline{\mathscr{F}}f) = T(f),$$

so every element of $\mathscr{S}'$ is the image under $\mathscr{F}$ of some element of $\mathscr{S}'$. An analogous argument holds for $\overline{\mathscr{F}}$. The equation (2–48) is an easy consequence of the definitions and (2–38):

$$(\mathscr{F}\overline{\mathscr{F}}T)(f) = (\overline{\mathscr{F}}T)(\mathscr{F}f) = T(\overline{\mathscr{F}}\mathscr{F}f) = T(f)$$

$$(\overline{\mathscr{F}}\mathscr{F}T)(f) = (\mathscr{F}T)(\overline{\mathscr{F}}f) = T(\mathscr{F}\overline{\mathscr{F}}f) = T(f). \quad \blacksquare$$

Theorem 2–2 suffices to justify almost all operations which will be carried out later, but the reader unacquainted with the subject is advised to prove some of the following statements directly from the definitions:

(a) $$\mathscr{F}[\delta(x)] = \frac{1}{(2\pi)^{n/2}}$$

(b) $$\mathscr{F}(\exp ik\cdot x) = (2\pi)^{n/2}\delta(p - k) \tag{2–49}$$

(c) Equations (2–35) and (2–36) are also valid for tempered distributions.

There is one other property of $\mathscr{F}$ which will be needed in the following: its relation to convolution. For functions in $\mathscr{S}$, say, we have

$$\begin{aligned}[\mathscr{F}(f * g)](k) &= (2\pi)^{-n/2}\int e^{-ik\cdot x}\,dx[\int f(x - \xi)g(\xi)\,d\xi] \\ &= (2\pi)^{-n/2}\int e^{-ik\cdot\xi}\,d\xi\,g(\xi)[\int e^{-ik\cdot(x-\xi)}\,d(x - \xi)f(x - \xi)] \\ &= (2\pi)^{n/2}(\mathscr{F}f)(k)(\mathscr{F}g)(k).\end{aligned} \tag{2–50}$$

So, apart from a factor $(2\pi)^{n/2}$, $\mathscr{F}$ converts a convolution into a product. This can be extended to the convolution of $f \in \mathscr{S}$ with $T \in \mathscr{S}'$:

$$\begin{aligned}[\mathscr{F}(f * T)](g) &= (f * T)(\mathscr{F}g) = T(\hat{f} * \mathscr{F}g) \\ &= (2\pi)^{n/2}T\{\mathscr{F}[(\overline{\mathscr{F}}\hat{f})g]\} = (2\pi)^{n/2}(\mathscr{F}T)[(\mathscr{F}f)g] \\ &= (2\pi)^{n/2}[(\mathscr{F}f)(\mathscr{F}T)](g),\end{aligned} \tag{2–51}$$

where we have used the definitions (2–46) and (2–31) of $\mathscr{F}$ and $f * T$ and the elementary identity $\overline{\mathscr{F}}\hat{f} = \mathscr{F}f$. We can write (2–51) as

$$f * T = (2\pi)^{n/2}\overline{\mathscr{F}}[(\mathscr{F}f)(\mathscr{F}T)], \tag{2–52}$$

in which case it displays the first part of

Theorem 2–3

The Fourier transform of a tempered distribution of fast decrease is an infinitely differentiable function which is bounded by a polynomial.

Proof:

It remains only to prove the polynomial boundedness. That follows since every tempered distribution is of the form (2–11).

2–3. LAPLACE TRANSFORMS AND HOLOMORPHIC FUNCTIONS

It is well known that an integral of the form

$$g(x) = \int_0^\infty e^{ikx}f(k)\,dk \tag{2–53}$$

where f, for concreteness sake, is in $\mathscr{S}(\mathbf{R}^1)$, is a boundary value of a holomorphic function, namely, the Laplace transform

$$g(x + iy) = \int_0^\infty e^{ik(x+iy)}f(k)\,dk \tag{2–54}$$

holomorphic in the upper half-plane, $y > 0$. [Clearly, the extra factor e^{-ky} in (2–54) only serves to improve the convergence of an already splendidly convergent integral. For $y > 0$,

$$\frac{dg(z)}{dz} = \lim_{\Delta z \to 0} \left[\frac{g(z + \Delta z) - g(z)}{\Delta z} \right]$$

exists and is independent of direction. That is one way to define what is meant by a holomorphic function of a single complex variable.] In this section, we present theorems which generalize this example in two respects. First, we admit several complex variables and replace the positive k axis by a convex cone. Second, we admit distributions rather than functions as boundary values. The generalization of the upper half-plane, $y > 0$, turns out to be a so-called tube in which the real parts of the complex variables are unrestricted but the imaginary parts are required to be in a cone. The functions holomorphic in the tube, which are Laplace transforms of tempered distributions vanishing outside a cone, are not arbitrary but possess certain characteristic boundedness properties. It is essential for applications to field theory to have these boundedness conditions spelled out in detail.

Since it is only in recent years that holomorphic functions of several complex variables have turned out to be of use in theoretical physics, we begin by recalling their definition and basic properties. We denote by $\mathbf{C}^n$ the n-dimensional vector space with complex scalars. A function f, defined in a neighborhood of a point w of $\mathbf{C}^n$, is said to be ***holomorphic*†** ***at the point*** w if there exists a multiple power series

$$\sum_{k_1 \ldots k_n = 0}^{\infty} a_{k_1 k_2 \ldots k_n} (z_1 - w_1)^{k_1} \ldots (z_n - w_n)^{k_n} \tag{2–55}$$

which converges for z in some neighborhood of w and is equal to $f(z)$ there. Just as in the case of power series in one variable, one proves that if (2–55) converges at z, it converges absolutely and uniformly for every ζ satisfying

$$|\zeta_j - w_j| \leqslant R_j = |z_j - w_j| - \varepsilon \qquad j = 1, \ldots n, \text{ for any } \varepsilon > 0 \tag{2–56}$$

We shall call such a region a *polydisc*.

There is a theorem of Weierstrass, familiar from the theory of holomorphic functions of one complex variable, which says that one can differentiate such a series term by term. Using this result one gets

$$a_{k_1 \ldots k_n} = \frac{1}{k_1! \ldots k_n!} \frac{\partial^{|k|} f(z_1, \ldots z_n)}{(\partial z_1)^{k_1} \ldots (\partial z_n)^{k_n}} \Bigg|_{z_1 = w_1, \ldots z_n = w_n} \tag{2–57}$$

† Often also termed *analytic*.

Thus, in particular, a holomorphic function of $z_1, \ldots z_n$ is an infinitely differentiable function of the n real parts of the z's if we take the imaginary parts equal to zero. Just how special it is can be seen more clearly if one uses Cauchy's integral formula n times to obtain

$$f(z_1, \ldots z_n) = \frac{1}{(2\pi i)^n} \int_{|\zeta_1 - w_1| = R_1, \ldots |\zeta_n - w_n| = R_n} \cdots \int \frac{f(\zeta_1, \ldots \zeta_n)\, d\zeta_1 \ldots d\zeta_n}{(\zeta_1 - z_1) \ldots (\zeta_n - z_n)} \tag{2–58}$$

as a representation valid for $z_1, \ldots z_n$ in a polydisc of uniform convergence of (2–55). From this there easily follows the estimate

$$\frac{1}{k_1! \ldots k_n!} \left| \frac{\partial^{|k|} f(z_1, \ldots z_n)}{(\partial z_1)^{k_1} \ldots (\partial z_n)^{k_n}} \right| \leqslant \frac{C}{R_1^{k_1} \ldots R_n^{k_n}}. \tag{2–59}$$

These inequalities are essentially all that distinguishes an infinitely differentiable function from a holomorphic function. In no physical context we know is (2–59) a direct physical statement as it stands, but it has been shown that in a number of physical situations, notably the ones considered in Chapters 3 and 4, the physical laws involved imply the inequalities (2–59) for important functions occurring in the theory. This has highly remarkable consequences as we shall see.

There are two important and related principles governing the behavior of holomorphic functions of a single complex variable which are also valid for holomorphic functions of several complex variables. *analytic continuation* and determination of a function by its values on a *real environment*. According to the first, one can extend the domain of a holomorphic function in a unique way by considering sequences of overlapping polydiscs. Of course, in general, using two different sequence of overlapping discs will lead to different values at the same point of $\mathbf{C}^n$, and one is forced to introduce a Riemann "surface" for the function in order to restore the single valuedness. This will not be necessary in any of the applications we have in mind. The second principle arises from the fact that all the coefficients of the power series for a holomorphic function can be obtained by calculating the derivatives in (2–57) in the direction of the real axis. Thus, a holomorphic function is determined in a full complex neighborhood of a point of $\mathbf{C}^n$ by its values on a real environment, that is, on an open set of $\mathbf{R}^n$ obtained by letting only the real parts of the complex variables vary.

An important example of the unique determination of a holomorphic function by its values on a real environment occurs in the study of sets of holomorphic functions with a transformation law under $L_+^\uparrow$ (see

Theorems 2–11 and 3–5). There we have a function defined on the restricted Lorentz group and want to assert that its extension to the complex Lorentz group is unique. This is an immediate consequence of what has been said above, provided we can find a parametrization of the Lorentz group such that the parameters, say $\lambda_1, \dots \lambda_6$, are real and independent for the real group, and complex and independent for the complex group, and the function in question is holomorphic in them. The parametrization need only work for a neighborhood, N, of the identity element because we can get a parametrization for a neighborhood of an arbitrary element, g, by multiplying all the elements of N by g and using the same parametrization for the neighborhood gN of g.

In the practical applications there will be two kinds of functions:

$$F(\Lambda\zeta_1, \dots \Lambda\zeta_n) \tag{2–60}$$

and

$$S(\Lambda)_{\alpha\beta},$$

where F is holomorphic in its arguments. Since a holomorphic function of a holomorphic function is holomorphic, the first function will be of the required kind if an appropriate parametrization of Λ has been found. The matrix $S(\Lambda)$ is one of the irreducible representations $\mathscr{D}^{(j/2,k/2)}$ of $SL(2,C)$, with $j + k$ even. [It is therefore a representation of $L_+^\uparrow$ because $\mathscr{D}^{(j/2,k/2)}(A) = \mathscr{D}^{(j/2,k/2)}(-A)$ in that case.] Now $\mathscr{D}^{(j/2,k/2)}(A)_{\alpha\beta}$ is a polynomial in the matrix elements of A, and their complex conjugates and the matrix elements of A can be expressed locally as analytic functions of the matrix elements of $\Lambda(A)$, so again parametrizing Λ properly leads to the desired expression. Incidentally, the analytic continuation is given by $\mathscr{D}^{(j/2,k/2)}(A, B)_{\alpha\beta}$, which can be expressed locally as a function of the matrix elements of $\Lambda(A, B)$.

Thus, it remains to give an appropriate parametrization of Λ. For this purpose, note that the matrix $\Lambda^\mu{}_\nu$ can be written as an exponential of a matrix $\Sigma^\mu{}_\nu$, and

$$\Lambda^T G \Lambda = G \qquad \text{is equivalent to } \Sigma^T = -G\Sigma G$$

or

$$\Sigma_{\mu\nu} = -\Sigma_{\nu\mu}.$$

The six parameters $\Sigma_{01}, \Sigma_{02}, \Sigma_{03}, \Sigma_{12}, \Sigma_{13}, \Sigma_{23}$ are arbitrary real for $L_+^\uparrow$ and arbitrary complex for $L_+(C)$, and can be taken for $\lambda_1, \dots \lambda_6$, provided one considers a sufficiently small neighborhood of the identity.

It is natural to ask whether the notion of real environment is also significant when it lies on the boundary of the domain in which a function is holomorphic. For one complex variable, that is well known to be the case; a generalization to n variables will be obtained in Theorem 2–17.

There is a useful criterion for a function of several variables to be holomorphic which we shall use several times in the following.

Theorem 2–4

Let F be a function defined in some open set D of $\mathbf{C}^n$. A necessary and sufficient condition that F be holomorphic is that it be continuous in all variables together and holomorphic in each variable separately.

Remark:

This theorem is also true if the condition of joint continuity is dropped. That is a deep result known as Hartog's theorem. (See Bochner and Martin, Chapter VII.) The analogue of Hartog's theorem for infinitely differentiable functions is false; this is another indication of the special character of holomorphic functions.

Proof:

In a suitable polydisc of center $w_1, \ldots w_n$ contained in D we can write (using the separate holomorphy) the successive integral

$$F(z_1, \ldots z_n) = \frac{1}{2\pi i} \int_{|\zeta_j - w_j| = R_j} \frac{d\zeta_1}{\zeta_1 - z_1} \frac{1}{2\pi i} \int \frac{d\zeta_2}{\zeta_2 - z_2} \cdots \frac{1}{2\pi i} \int \frac{d\zeta_n}{\zeta_n - z_n} F(\zeta_1, \ldots \zeta_n).$$

But, because of the joint continuity, the successive integral can be written as a multiple integral. Writing the series expansion

$$\frac{1}{(\zeta_1 - z_1) \ldots (\zeta_n - z_n)} = \sum_{k_1 \ldots k_n = 0}^{\infty} \frac{(z_1 - w_1)^{k_1}}{(\zeta_1 - w_1)^{k_1+1}} \cdots \frac{(z_n - w_n)^{k_n}}{(\zeta_n - w_n)^{k_n+1}},$$

which converges uniformly in every subpolydisc, and interchanging the order of integration and summation, we get a convergent series expansion for F. ∎

Like every continuous function, a holomorphic function, F, defined on an open set $\mathcal{O}$ is a distribution in $\mathcal{D}(\mathcal{O})'$:

$$F(f) = \int dx_1\, dy_1 \ldots dx_n\, dy_n\, f(x_1, y_1, \ldots x_n, y_n) F(z_1, \ldots z_n),$$

where f is any test function with support lying in $\mathcal{O}$. In the proof of the edge of the wedge theorem in Section 2–5 we shall need to know the relation between convergence in $\mathcal{D}(\mathcal{O})'$ and uniform convergence on compact sets for a sequence of holomorphic functions, F_k, $k = 1, 2, \ldots$. The answer is simple: the two notions are identical. Clearly, since a sequence of holomorphic F_k converging uniformly on compact subsets of $\mathcal{O}$ converges to a function, F, holomorphic in $\mathcal{O}$, we have

$$F_k(f) \to F(f) \tag{2-61}$$

for $f \in \mathcal{D}(\mathcal{O})$. Conversely, if $\lim_{k\to\infty} F_k(f)$ exists for every $f \in \mathcal{D}(\mathcal{O})$, we can smear the multiple Cauchy formula

$$F_k(w_1, \ldots w_n) = \frac{1}{(2\pi)^n} \int_0^{2\pi} d\theta_1 \ldots \int_0^{2\pi} d\theta_n F_k(w_1 + R_1 e^{i\theta_1}, \ldots w_n + R_n e^{i\theta_n})$$

in the radii $R_1, \ldots R_n$ with an infinitely differentiable function g with support lying on the product of the positive real axes and sufficiently near zero, and such that

$$\int_0^\infty \cdots \int_0^\infty R_1\, dR_1 \ldots R_n\, dR_n\, g(R_1, \ldots R_n) = 1.$$

Then

$$F_k(w_1, \ldots w_n) = F_k(f_w),$$

where

$$f_w(x_1, y_1, \ldots x_n, y_n) = (2\pi)^{-n} g(|z_1 - w_1|, \ldots |z_n - w_n|).$$

Now $f_w \in \mathcal{D}(\mathcal{O})$, so convergence in $\mathcal{D}(\mathcal{O})'$ implies the convergence of the $F_k(w_1, \ldots w_n)$ at each point $w_1, \ldots w_n \in \mathcal{O}$. Furthermore, since, as $w_1, \ldots w_n$ varies over a sufficiently small compact set, K, f_w varies over a bounded set in $\mathcal{D}(\mathcal{O})$, the convergence is uniform on K. The limit function F is therefore a holomorphic function. This completes our general remarks about holomorphic functions. We now turn to the definition and properties of the Laplace transform.

If T is a distribution in $\mathcal{D}'_p$, it may happen that $e^{-p\cdot\eta}T$, which is certainly a distribution in $\mathcal{D}'$, is also a distribution in $\mathcal{S}'$. In this case, we can define the *Laplace* transform of T as the distribution of $\mathcal{S}'_\xi$ given by

$$\mathcal{L}(T) = \mathcal{F}(e^{-p\cdot\eta}T). \tag{2-62}$$

$\mathcal{L}(T)$ actually depends parametrically on η but we shall not indicate that explicitly. For a function T, this definition (2–62) is

$$\mathcal{L}(T)(\xi, \eta) = \frac{1}{(2\pi)^{n/2}} \int e^{-ip\cdot(\xi - i\eta)} T(p)\, dp. \tag{2-63}$$

The convention of signs in the Fourier transform which is used here leads to $\xi - i\eta$ rather than $\xi + i\eta$ in the exponent. This results from our effort to agree with customary physical notation in Chapters 3 and 4. T need not have any special support properties for this definition to make sense. For example, if $T(p) = \exp(-|p|^2)$ the Laplace transform certainly exists for all η. Thus, we are dealing with what is sometimes called the two-sided Laplace transform, which includes the one-sided transform (2–53) as a special case.

The first fact about the η's for which $\mathscr{L}(T)$ exists is given by Theorem 2–5.

Theorem 2–5

Let T be a distribution of $\mathscr{D}'_p$. The set of all η such that $e^{-p\cdot\eta}T$ is in $\mathscr{S}'_p$ is convex.

Proof:

Suppose $e^{-p\cdot\eta'}T$ and $e^{-p\cdot\eta''}T$ belong to $\mathscr{S}'_p$. Let $\eta = t\eta' + (1-t)\eta''$ with $0 \leqslant t \leqslant 1$. Define the infinitely differentiable function a by

$$a(p) = \exp(-p\cdot\eta)[\exp(-p\cdot\eta') + \exp(-p\cdot\eta'')]^{-1}. \tag{2–64}$$

Then a is bounded. [For any two real positive numbers, $c < d$ implies $c^t < d^t$ and therefore $c < c^t d^{(1-t)} < d$. Applying this to $e^{-p\cdot\eta'}$ and $e^{-p\cdot\eta''}$ gives $0 < a(p) \leqslant 1$.] Furthermore, all the partial derivatives of a with respect to p are bounded. This follows because they are either linear combinations of products of the form (2–64) or analogous functions in which η is replaced by η' or η''. Consequently, the identity

$$\exp(-p\cdot\eta)T = a[\exp(-p\cdot\eta')T] + a[\exp(-p\cdot\eta'')T]$$

displays $\exp(-p\cdot\eta)T$ as the sum of two distributions in $\mathscr{S}'_p$. ∎

Theorem 2–5 shows that the definition (2–62) of the Laplace transform associates with every $T \in \mathscr{D}'$ a convex set of η (possibly empty!) for which the Laplace transform exists. Now we take a given set, Γ, of η's and study what distributions T can have Laplace transforms defined for $\eta \in \Gamma$.

Theorem 2–6

Let Γ be a convex open set of $\mathbf{R}^n$ and T a distribution $\in \mathscr{D}'_p$ such that $e^{-p\cdot\eta}T \in \mathscr{S}'_p$ for all $\eta \in \Gamma$. Then $\mathscr{L}(T)$ is a holomorphic function of $\xi - i\eta$ for all $\xi - i\eta$ in the tube $\mathbf{R}^n - i\Gamma$. $\mathscr{L}(T)$ satisfies the boundedness condition

$$|\mathscr{L}(T)(\xi - i\eta)| \leqslant |P_K(\xi)| \tag{2–65}$$

for some polynomial P_K when η varies over any compact subset, K, of Γ.

Conversely, every function holomorphic in the tube $\mathbf{R}^n - i\Gamma$, and satisfying (2–65) for some polynomial P_K for each compact subset, K, of Γ, is the Laplace transform of a uniquely determined distribution $T \in \mathscr{D}'_p$, such that $e^{-p\cdot\eta}T \in \mathscr{S}'_p$ for all $\eta \in \Gamma$.

Remarks:

1. A tube clearly is a subset of $\mathbf{C}^n$ with a special property of translation invariance: it is invariant under real translations. In the mathematical literature the notation is customarily arranged so that the invariance is under purely imaginary translations instead. This evidently is a matter of convention.

2. In the case of one complex variable, $\xi - i\eta$, and Γ, the cone $\eta > 0$, the theorem deals with holomorphic functions in the lower half-plane. Then the boundedness restriction (2–65) asserts that the holomorphic function is bounded by a polynomial $P_{a,b}(\xi)$ in every horizontal strip of the form $0 < a \leqslant \eta \leqslant b < \infty$. Since the polynomial may vary with a and b we have no control over the behavior of the holomorphic function for large η and small η. The following two examples illustrate this point.

$$e^{-z^2/2} = \frac{1}{\sqrt{2\pi}} \int_{-\infty}^{\infty} dk\, e^{ikz} e^{-k^2/2} \tag{2–66}$$

and

$$\sqrt{\pi}\,\frac{e^{i/4z}}{\sqrt{-iz}} = \int_0^{\infty} dk\, e^{ikz}\,\frac{\cosh\sqrt{k}}{\sqrt{k}}. \tag{2–67}$$

The first grows as $e^{y^2/2}$ along the imaginary axis as $y \to \infty$, while the second grows as $y^{-1/2}e^{1/4y}$ as $y \to 0$.

Proof:

Our first step is to show that for η varying in certain subsets of Γ, $e^{-p\cdot\eta}T$ can be written as a linear combination of distributions in $\mathscr{S}'_p$, each of which is a product of a function in $\mathscr{S}$ with a distribution in $\mathscr{S}'$; that is, $e^{-p\cdot\eta}T$ is a tempered distribution of fast decrease. The Fourier transform of such a distribution is an infinitely differentiable function of ξ according to Theorem 2–3. The particular form will also guarantee that it is infinitely differentiable in ξ and η together. Then we shall show that this function satisfies the Cauchy–Riemann equations in ξ and $-\eta$ and is therefore a holomorphic function of $\xi - i\eta$ according to Theorem 2–4.

Let $\eta^{(1)}, \ldots \eta^{(\ell)}$ be a set of vectors of Γ such that their convex hull, H, that is, the cone of vectors $\sum_{j=1}^{\ell} t_j \eta^{(j)}$, for $\sum_{j=1}^{\ell} t_j = 1, t_j \geqslant 0, j = 1, \ldots l$, has a non-empty interior. If η is a vector of that interior, then the function

$$a(p, \eta) = \exp(-p\cdot\eta)\left[\sum_{j=1}^{\ell} \exp(-p\cdot\eta^{(j)})\right]^{-1} \tag{2–68}$$

is bounded in p. [The argument is quite analogous to that for the boundedness of (2–64): If $a_i > 0$ then

$$\textstyle\sum_{i=1}^{\ell} t_i \log a_i \leqslant \sup_i \log a_i \qquad \text{for} \quad \sum_{i=1}^{\ell} t_i = 1,\ t_i \geqslant 0,\ i = 1 \ldots \ell.$$

Therefore $|a(p, \eta)| \leqslant 1$.] Furthermore, the same holds for all the derivatives with respect to p, and uniformly provided η stays in H. The derivatives of $a(p, \eta)$ with respect to p and η are bounded by polynomials in p uniformly for η in H. Thus, writing

$$\exp(-p\cdot\eta)T = \sum_{j=1}^{\ell} a(p, \eta)[\exp(-p\cdot\eta^{(j)})T] \tag{2–69}$$

we display the left-hand side as a linear combination of elements of $\mathscr{S}'_p$, namely, the square brackets, times an infinitely differentiable function of p, all of whose derivatives with respect to p are bounded and whose derivatives with respect to η are polynomial bounded. This is already enough to show that the Fourier transform of $\exp(-p\cdot\eta)T$ is a distribution in $\mathscr{S}'_p$ which is differentiable in η; but to get that it is a function we need a stronger statement than (2–69).

We assert that for any compact subset, K, of Γ, there is an $\varepsilon > 0$ such that

$$\exp(\varepsilon\sqrt{1 + |p|^2}) \exp(-p\cdot\eta)T = T_1 \tag{2–70}$$

$\in \mathscr{S}'_p$ for all $\eta \in K$. To prove this we pick an $\eta \in K$. Then, if ρ is a real vector satisfying $|\rho| \leqslant \varepsilon$, the set $\eta + \rho$ will lie in the convex hull of a finite number of vectors of Γ, for sufficiently small ε. Thus, for these $\eta + \rho$'s, one can construct $a(p, \eta + \rho)$ as in (2–68). If we define

$$b(p, q) = \exp[\varepsilon\sqrt{1 + |p|^2}]a(p, q)$$

for $|q - \eta| \leqslant \varepsilon$ we have

$$\begin{aligned}|b(p, q)| &\leqslant \exp[\varepsilon(1 + |p|)]|a(p, q)| \\ &\leqslant \exp(\varepsilon) \sup_{|\rho| \leqslant \varepsilon} \exp(-\rho \cdot p)|a(p, q)| \\ &\leqslant \exp(\varepsilon) \sup_{|\rho| \leqslant \varepsilon} |a(p, q + \rho)|.\end{aligned}$$

Thus $b(p, q)$ is bounded in p for all q sufficiently close to η. Furthermore, its derivatives with respect to p and q are bounded by polynomials in p, since they are linear combinations of derivatives of $\exp(\varepsilon\sqrt{1 + |p|^2})$ and those of $a(p, q)$. If we write

$$\exp(\varepsilon\sqrt{1 + |p|^2}) \exp(-p \cdot \eta)T = \sum_{j=1}^{\ell} b(p, \eta)[\exp(-p \cdot \eta^{(j)})T]$$

we have (2–70) only for η in the neighborhood of a point of K. But, since K is compact it can be covered by a finite number of such neighborhoods, and therefore the assertion is valid throughout K if the least of the ε is taken. The validity of (2–70) assures us that for each fixed compact, K, $\exp(-p \cdot \eta)T = \exp(-\varepsilon\sqrt{1 + |p|^2})T_1$, where T_1 is in $\mathscr{S}'_p$, and varies over a bounded set when η varies over K. Thus, $\exp(-p \cdot \eta)T$ is a tempered distribution of fast decrease as promised, and $\mathscr{L}(T)(\xi, \eta)$ is an infinitely differentiable function of ξ and η. Furthermore, for $\eta \in K$,

$$|\mathscr{L}(T)(\xi, \eta)| < |P_K(\xi)|, \tag{2–71}$$

where P_K is some polynomial.

Now

$$\begin{aligned}\frac{\partial}{\partial \xi_j} \mathscr{L}(T)(\xi, \eta) &= \mathscr{F}(-ip_j e^{-p \cdot \eta} T) \\ &= i \frac{\partial}{\partial \eta_j} \mathscr{L}(T)(\xi, \eta)\end{aligned}$$

which are the Cauchy–Riemann equations for the complex variables $\xi_j - i\eta_j$, $j = 1, \ldots n$. Therefore $\mathscr{L}(T)$ is a holomorphic function of $\xi - i\eta$. This completes the proof of the first half of the theorem.

Conversely, suppose $F(\xi - i\eta)$ is a holomorphic function in the tube $\mathbf{R}^n - i\Gamma$, where Γ is an open convex set, and suppose that for each compact subset K of Γ there exists a polynomial P_K such that

$$|F(\xi - i\eta)| \leqslant |P_K(\xi)| \qquad \eta \in K. \tag{2–72}$$

Equation (2–72) permits us to say that, for each $\eta \in \Gamma$, $F(\xi - i\eta)$ is a distribution in ξ so we can immediately write down the distribution, $\hat{F} \in \mathscr{S}'_p$, which, when multiplied by $e^{\eta \cdot p}$, ought to be that of which F is the Laplace transform

$$\hat{F}(p, \eta) = \overline{\mathscr{F}}_\xi[F(\xi - i\eta)].$$

If we multiply it by $e^{p \cdot \eta}$ we get a distribution which may no longer be in $\mathscr{S}'_p$ but must be in $\mathscr{D}'_p$. If we could show that

$$\frac{\partial}{\partial \eta_j}[e^{p \cdot \eta}\hat{F}(p, \eta)] = 0 \tag{2–73}$$

we would write

$$\hat{F}(p, \eta) = e^{-p \cdot \eta}T(p) \qquad T \in \mathscr{D}'_p \tag{2–74}$$

and the proof would be complete. However, at the present stage we do not even know that $\hat{F}(p, \eta)$ is differentiable in η, let alone that (2–73) holds. To get the differentiability we show that the holomorphy of F and the inequality (2–72) imply analogous inequalities for the derivatives of F.

We first explain the idea of the proof for the case of one variable. Then we take as the compact set inside the tube a closed interval $0 < a \leqslant \eta \leqslant b$. (Its position in the lower half-plane of $\xi - i\eta = z$ can always be arranged by suitable location of the origin of coordinates.) We choose a sufficiently high power, z^k, of z so that $f(z)z^{-k}$ is bounded in the strip $a \leqslant \eta \leqslant b$. Then, with the contour C, shown in Figure 2–1, we obtain the integral representations for $a < \operatorname{Im} z < b$

$$\begin{aligned} \frac{f(z)}{z^{k+2}} &= \frac{1}{2\pi i}\int_C \frac{f(\zeta)\, d\zeta}{\zeta^{k+2}(\zeta - z)} \\ \frac{df/dz}{z^{k+2}} &= (k + 2)\frac{f(z)}{z^{k+3}} + \frac{1}{2\pi i}\int_C \frac{f(\zeta)\, d\zeta}{\zeta^{k+2}(\zeta - z)^2}. \end{aligned} \tag{2–75}$$

From these representations one immediately derives that $|f(z)| \leqslant c|z|^k$ in the strip implies $|df/dz| \leqslant d|z|^{k+2}$. For n variables the argument is similar. One takes a compact subset of Γ which is a product of n intervals and argues that $F(z)(z_1^{-k_1}\ldots z_n^{-k_n})$ will be bounded for sufficiently large $k_1\ldots k_n$ by virtue of (2–72). (Again, by suitable choice of origin, the intervals have been taken in the lower half-plane.) Instead of (2–75) one has a multiple Cauchy integral formula, but the result is the same: the polynomial boundedness of the partial derivatives of $F(\xi - i\eta)$.

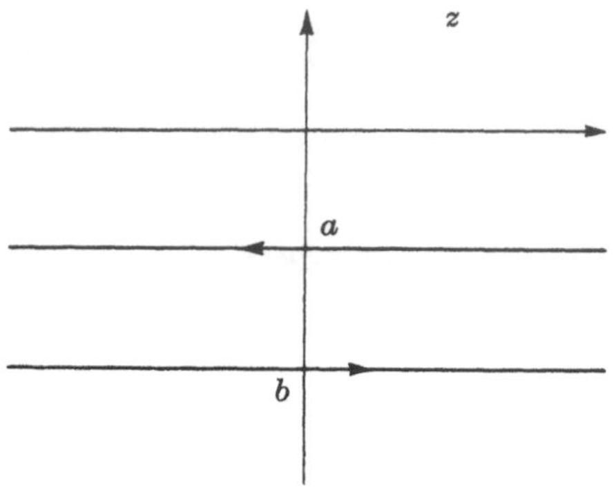

FIGURE 2–1. The contour C in the plane of $z = \xi - i\eta$.

Since that in turn implies the differentiability of $\hat{F}$, we can carry out the differentiations in (2–73):

$$\frac{\partial}{\partial\eta_j}[e^{p\cdot\eta}\hat{F}(p,\eta)] = p_j e^{p\cdot\eta}\hat{F}(p,\eta) + e^{p\cdot\eta}\frac{\partial\hat{F}}{\partial\eta_j}(p,\eta)$$

$$= e^{p\cdot\eta}\mathscr{F}\left[i\frac{\partial}{\partial\xi_j}F(\xi - i\eta) + \frac{\partial F}{\partial\eta_j}(\xi - i\eta)\right] = 0. \quad \blacksquare$$

We now turn to the further restrictions on the holomorphic functions $\mathscr{L}(T)$ which result from the requirements that T have support in a cone and that T be tempered.

First consider the matter heuristically. If the support of T lies in a half-space $p\cdot a > A$, where a is some fixed vector, then, for $\eta \in \Gamma$, the "integral"

$$\int e^{-ip\cdot(\xi - i\eta)}T(p)\,dp \tag{2–76}$$

converges even better if the exponential is replaced by

$$e^{-ip\cdot(\xi - i\eta)}e^{-t(p\cdot a - B)} \qquad \text{for } B < A \qquad t \geqslant 0$$

i.e.,

$$e^{tB}\int e^{-ip\cdot[\xi - i(\eta + ta)]}T(p)\,dp$$

ought to exist for all $B < A$ and $t \geqslant 0$. But that means that if $\eta \in \Gamma$ then $\eta + ta \in \Gamma$ for all $t \geqslant 0$ when a is any vector such that the support of T lies in the half-space $p\cdot a > A$. Of course, this statement is empty unless Γ is non-empty. If T is tempered, then Γ contains at least one point $\eta = 0$, because for $\eta = 0$, the Laplace transform is the Fourier transform. It need not have any other point; consider, for example, $T(p) = (1 + |p|^2)^{-2n}$. However, if T is tempered *and* the support of T lies in a half-space, the combination of the preceding heuristic discussion and the fact that Γ must be convex forces Γ to be a cone. That is the situation met in quantum field theory as we shall see.

Theorem 2–7

Suppose $T \in \mathscr{D}'_p$ and Γ (a convex set of $\mathbf{R}^n$) consists of vectors η such that $e^{-p\cdot\eta}T \in \mathscr{S}'_p$. If T has its support lying in a half-space $p \cdot a > A$, then Γ contains all points of the form $\eta + ta$ with $\eta \in \Gamma$ and $t \geqslant 0$.

Proof:

Let $\varepsilon > 0$ and choose an infinitely differentiable function q of one variable so that it is 1 for $x \geqslant A$ and 0 for $x \leqslant A - \varepsilon$. Define $g_\varepsilon(p) = q(p \cdot a)$. The distribution $\exp[-p\cdot(\eta + ta) + At]T$ is in $\mathscr{D}'$ for $\eta \in \Gamma$, and has support in the half-space $p \cdot a \geqslant A$ because T does. Furthermore, if $f \in \mathscr{D}$, $t \geqslant 0$,

$$\{\exp[-p\cdot(\eta + ta) + At]T\}(f) = [\exp(-p\cdot\eta)T][\exp(-tp\cdot a + At)]g_\varepsilon f\} \quad (2\text{–}77)$$

is independent of g_ε provided it satisfies the above requirements for some ε. Now the right-hand side of (2–77) has an extension to $\mathscr{S}'$ by continuity because $[\exp(-tp\cdot a + At)]g_\varepsilon f$ is in $\mathscr{S}$ if f is, and depends continuously on f. Therefore, $\exp[-p\cdot(\eta + ta)T] \in \mathscr{S}'_p$, so $\eta + ta \in \Gamma$. ∎

In the applications of Chapters 3 and 4, T is defined for arguments $p_1, \ldots p_n$ where each p_j is a real four-vector, and vanishes unless each p_j lies in or on the plus light cone. This phrase will occur so frequently in the following that we introduce the notation V_+ to stand for all real four-vectors p, satisfying $p^2 = (p^0)^2 - (\mathbf{p})^2 > 0$ and $p^0 > 0$, and $\overline{V}_+$ to stand for the closure of V_+, the set of p satisfying $p^2 \geqslant 0, p^0 \geqslant 0$. Since, in Chapters 3 and 4, T will also be tempered, application of Theorem 2–7 yields that $\mathscr{L}(T)(\xi - ia)$ is analytic for all a of the form $a_1 \ldots a_n$ with $a_j \in V_+, j = 1, \ldots n$. In the following, the tube $\mathbf{R}^n - i\Gamma$, where $\Gamma = (a_1, \ldots a_n)$ with $a_j \in V_+$, $j = 1, \ldots n$, will be denoted $\mathscr{T}_n$, and sometimes called *the tube* without further specification.

The Laplace transforms of distributions with support in a half-space possess stronger boundedness properties than (2–72). While these can be proved for general tubes, we get an especially neat statement for $\mathscr{T}_n$ so we confine our attention to that case.

Theorem 2–8

Let $T \in \mathscr{D}'_p$ and $e^{-p\cdot\eta}T \in \mathscr{S}'_p$ for all $\eta \in \Gamma$. Here $p\cdot\eta$ stands for $\sum_{j=1}^{n} \sum_{\mu=0}^{3} p_{j\mu}\eta_j^{\mu}$ and Γ for the cone $\eta_j \in V_+$, $j = 1,\ldots n$. Suppose $p \in \operatorname{supp} T$ implies $p_j \in \overline{V}_+$, $j = 1,\ldots n$. Then, for each $\eta \in \Gamma$, there is a polynomial P_η such that

$$|\mathscr{L}(T)[\xi - i(\eta + a)]| \leqslant |P_\eta(\xi - ia)| \qquad (2\text{–}78)$$

for all ξ and all $a \in \Gamma$.

For the converse, if F is a function holomorphic in $\mathscr{T}_n = \mathbf{R}^{4n} - i\Gamma$ and satisfying the inequality (2–78) for each $\eta \in \Gamma$ and some polynomial P_η, then F is the Laplace transform of a distribution with support in Γ.

Proof:

The proof of (2–78) uses the Bros–Epstein–Glaser lemma. See pages 21–27 in M. Reed and B. Simon, *Methods of Modern Mathematical Physics II, Fourier Analysis, Self-Adjointness*, Academic Press, 1975.

Conversely, suppose F is holomorphic in $\mathbf{R}^{4n} - i\Gamma = \mathscr{T}_n$ and is bounded according to (2–78). Then it certainly satisfies (2–72), so it is the Laplace transform of a distribution $T \in \mathscr{D}'$ such that $e^{-p\cdot\eta}T \in \mathscr{S}'_p$ for all $\eta \in \Gamma$. It remains to show that the support of T lies inside Γ.

Pick a test function g of compact support lying entirely outside Γ. Then

$$\begin{aligned} T(g) &= [e^{-p\cdot(\eta+a)}T][e^{p\cdot(\eta+a)}g] = \{\mathscr{F}[e^{-p\cdot(\eta+a)}T]\}\{\overline{\mathscr{F}}[e^{p\cdot(\eta+a)}g]\} \\ &= \int d\xi\, F[\xi - i(\eta+a)]G[\xi - i(\eta + a)]\, d\xi, \end{aligned}$$

where

$$G(\xi - i\eta) = (2\pi)^{-2n} \int e^{ip\cdot(\xi - i\eta)}g(p)\, dp.$$

Now, by appropriate choices of a it can be arranged that, throughout the support of g, we have $p\cdot(\eta + a) < 0$, and as $a \to \infty$, $p\cdot(\eta + a) \to -\infty$. Then, as $a \to \infty$ in Γ, $G[\xi - i(\eta + a)] \to 0$ in $\mathscr{S}_\xi$. Therefore

$$\begin{aligned} |T(g)| &\leqslant \int d\xi\, |P_\eta(\xi - ia)|\, |G[\xi - i(\eta + a)]| \\ &\leqslant \int \frac{d\rho}{[1 + |\rho|^2]^k} \sup_\xi [1 + |\xi|^2]^k |P_\eta(\xi - ia)G[\xi - i(\eta + a)]| \to 0 \end{aligned}$$

as $a \to \infty$ for sufficiently large fixed k. ∎

It is already visible in the example (2–67) that a holomorphic function which is a Laplace transform need not have a boundary value even in the sense of distribution theory. If it is the Laplace transform of a tempered distribution, T, however, the boundary value is just the Fourier transform of T.

Theorem 2–9

Suppose $T \in \mathscr{S}'_p$ and $\mathscr{L}(T)$ exists for all $\eta \in \Gamma$ as described in Theorem 2–8. Then

$$\lim_{\eta \to 0} \int \mathscr{L}(T)(\xi - i\eta) f(\xi)\, d\xi = [\mathscr{F}(T)](f) \qquad (2\text{–}79)$$

that is, $\mathscr{L}(T)$ converges in $\mathscr{S}'_\xi$ to $\mathscr{F}(T)$ as $\eta \to 0$ inside any closed cone in Γ.

Conversely, if $\mathscr{L}(T)$ converges in $\mathscr{S}'_\xi$ as $\eta \to 0$ in any such cone, T is a tempered distribution.

Proof:

The first statement of the theorem follows directly from (2–69) and the argument given in connection with it; $\exp(-p \cdot \eta)T$ is a continuous function of η with values in $\mathscr{S}'_p$ in the convex hull of any finite set of vectors of Γ.

The converse statement means that

$$\left[\lim_{\eta \to 0} \mathscr{F}(e^{-p \cdot \eta} T)\right](f) = \lim_{\eta \to 0} T(e^{-p \cdot \eta} \mathscr{F} f)$$

exists for every $f \in \mathscr{S}$. The completeness of $\mathscr{S}'$ implies there exists a tempered distribution T_1 such that this limit is $T_1(\mathscr{F}f)$. But for $\mathscr{F}f \in \mathscr{D}$, clearly $T_1(\mathscr{F}f) = T(\mathscr{F}f)$. Therefore $T_1 = T$ and T is tempered. ∎

Theorem 2–9 gives a characterization of those holomorphic functions which arise by Laplace transformation from tempered distributions but it is not expressed very directly in terms of $\mathscr{L}(T)$ as a holomorphic function. The last theorem of this section gives a more direct statement.

Theorem 2–10

Let $e^{-p \cdot \eta} T \in \mathscr{S}'_p$ for $\eta = 0$ and $\eta \in \Gamma$, with Γ as described in Theorem 2–8. Then, for each compact subset K of Γ, there is a polynomial P_K and an integer r such that

$$|\mathscr{L}(T)(\xi - it\eta)| \leqslant \frac{P_K(\xi)}{t^r} \qquad (2\text{–}80)$$

for all ξ, all t with $0 < t < 1$, and all $\eta \in K$.

Conversely, if F is a function holomorphic in $\mathscr{T}_n$ and satisfying (2–80), $F = \mathscr{L}(T)$, where T is a tempered distribution.

Proof:

To get (2–80), we repeat part of the argument used in the proof of Theorem 2–6. It shows that we can write $e^{-tp\cdot\eta}T$ as $a(p, t\eta)T_1$, where $T_1 \in \mathscr{S}'$ and $a \in \mathscr{S}_p$ for all $\eta \in K$, $0 < t < 1$ and a varies over a bounded set as η varies. Therefore, $\mathscr{L}(T)(\xi - it\eta)$ is the tempered distribution T_1 evaluated for the test function $(2\pi)^{-2n}a(p, t\eta)e^{-ip\cdot\xi}$. Consequently, for some integers k, l and some constant C

$$|\mathscr{L}(T)(\xi - it\eta)| \leqslant C\|e^{-ip\cdot\xi}a(p, t\eta)\|_{k,l} \tag{2–81}$$

A straightforward estimate shows that the right-hand side is $\leqslant t^{-r}P_k(\xi)$ for some integer r as η varies over K.

To verify the converse statement, we choose $f \in \mathscr{S}$ and study

$$\int F(\xi - it\eta)f(\xi)\, d\xi = h(t)$$

as a function of t. We have

$$\frac{dh}{dt}(t) = \int \sum_j \frac{\partial}{\partial(\xi - it\eta)_j} F(\xi - it\eta)(-i\eta_j)f(\xi)\, d\xi$$

$$= \int F(\xi - it\eta)i\eta\cdot\frac{\partial}{\partial\xi}f(\xi)\, d\xi$$

$$\vdots$$

$$h^{(j)}(t) = \int F(\xi - it\eta)\left(i\eta\cdot\frac{\partial}{\partial\xi}\right)^j f(\xi)\, d\xi.$$

Thus, by (2–80),

$$|h^{(j)}(t)| \leqslant C \sup_{\xi} \frac{|P_n(\xi)| \left|\left(i\eta\cdot\frac{\partial}{\partial\xi}\right)^j f(\xi)\right|(1 + |\xi|^2)^k}{t^r} \tag{2–82}$$

for some fixed sufficiently large k.

Now

$$h^{(j)}(t) = -\int_t^1 d\tau\, h^{(j+1)}(\tau) + h^{(j)}(1)$$

so, for $r > 1$,

$$|h^{(j)}(t)| \leqslant \left(\frac{1}{t^{r-1}} - 1\right)\frac{E}{r-1} + |h^{(j)}(1)|, \tag{2–83}$$

where E stands for the numerator in (2–82). This last equation shows that $h^{(j)}(t)$ is bounded in t by a sum of terms, each of which goes to zero as $f \to 0$ in $\mathscr{S}$, and of degree $-(r - 1)$ in t. We insert this expression in the formula for $h^{(j-1)}(t)$ and get an analogous statement with degree $-(r - 1)$ replaced by $-(r - 2)$. If we begin the process with a j sufficiently large, say $j = r + 1$, and generalize (2–83) to cover the case in which $1/\tau$ has to be integrated, we see that $h(t)$ is bounded in t by a sum of terms which approach zero as $f \to 0$ in $\mathscr{S}$. Therefore

$$\lim_{|t \to 0|} \int F(\xi - it\eta)f(\xi)\, d\xi$$

exists and, by Theorems 2–6 and 2–9, F is the Laplace transform of a tempered distribution.

We can roughly summarize the results of Theorems 2–7 through 2–10: the holomorphic functions which are Laplace transforms of tempered distributions with supports in Γ are holomorphic in the variable $\xi - i\eta$ in $\mathscr{T}_n$, and of polynomial growth in η near infinity and zero.

2–4. TUBES AND EXTENDED TUBES

We have seen in the preceding section that the Laplace transform of a tempered distribution vanishing outside a cone is the boundary value of a function holomorphic in a certain tube. In the present section we consider a function or a set of functions holomorphic in that tube and with a definite transformation law under $SL(2,C)$ or, what turns out to be the same thing, the restricted Lorentz group $L_+^\uparrow$. We show that these functions are necessarily holomorphic in a larger domain, the so-called extended tube, and satisfy a transformation law under the proper complex Lorentz group, $L_+(C)$.

As we have defined it, the tube $\mathscr{T}_n$ is the open set of complex $4n$ space, $\mathbf{C}^{4n}$, given by $\eta_j \in V_+$, $j = 1, \ldots n$, where $\zeta_j = \xi_j - i\eta_j$, $j = 1, \ldots n$ is the splitting of ζ_j into real and imaginary parts. The *extended tube* $\mathscr{T}_n'$ is the union of the open sets obtained from $\mathscr{T}_n$ by applying all proper complex Lorentz transformations. In other words, $\zeta_1, \ldots \zeta_n \in \mathscr{T}_n'$, if and only if there exists a $\Lambda \in L_+(C)$ and a point $w_1, \ldots w_n \in \mathscr{T}_n$ such that

$$\zeta_1, \ldots \zeta_n = \Lambda w_1, \ldots \Lambda w_n.$$

The transformation laws of sets of holomorphic functions with which we are concerned are of the form

$$\sum_{\beta} S(A)_{\alpha\beta} f_\beta(\zeta_1, \ldots \zeta_n) = f_\alpha(\Lambda(A)\zeta_1, \ldots \Lambda(A)\zeta_n), \qquad (2\text{–}84)$$

where $A \to S(A)$ is a matrix representation of $SL(2,C)$. In (2–84), the point $(\zeta_1, \ldots \zeta_n)$ lies in $\mathscr{T}_n$. We noted in Section 1–3, after (1–25), that by choosing suitable linear combinations of the f_β we can obtain equivalent equations for the linear combinations which transform according to irreducible representations of $SL(2,C)$, so it is no loss in generality to assume from the outset that $A \to S(A)$ is an irreducible representation, say $\mathscr{D}^{(j/2, k/2)}$. We conclude immediately that $j + k$ is even if the f_α are not all zero, because (2–84) evaluated for $A = -1$ reads, according to (1–26),

$$(-1)^{j+k} f_\alpha(\zeta_1, \ldots \zeta_n) = f_\alpha(\zeta_1, \ldots \zeta_n).$$

This implies that $A \to S(A)$ is in fact a representation of $L_+^\uparrow(R)$, since, if $j + k$ is even, $S(A) = S(-A)$.

In (2–84) let us fix $\zeta_1, \ldots \zeta_n \in \mathscr{T}_n$ and consider both sides as functions of the six real parameters of the group $SL(2,C)$, suitably chosen, or equivalently, the six real parameters of the Lorentz group $L_+^\uparrow$. The matrix $S(A) = S[\Lambda(A)]$ is an analytic function of the parameters in Λ, defined for Λ real, and so possesses the unique analytic continuation, $S(A, B)$ say, to complex Lorentz transformations $\Lambda(A, B)$ in some neighborhood of the set of real $\Lambda \in L_+^\uparrow$. Here and in the following, a neighborhood of a complex Lorentz transformation Λ_1 will mean all $\Lambda \in L_+(C)$ such that with suitable parametrization the six parameters of Λ lie in a complex neighborhood of the six parameters defining Λ_1. A neighborhood of a set of Lorentz transformations is defined analogously. The analytic continuation of the right-hand side of (2–84) is $f_\alpha[\Lambda(A, B)\zeta_1, \ldots \Lambda(A, B)\zeta_n]$; thus the analytically continued equation is

$$\sum_\beta S(A, B)_{\alpha\beta} f_\beta(\zeta_1, \ldots \zeta_n) = f_\alpha[\Lambda(A, B)\zeta_1, \ldots \Lambda(A, B)\zeta_n], \quad (2\text{–}85)$$

which holds in some complex neighborhood of any real $\Lambda \in L_+^\uparrow$, in particular, in some complex neighborhood of the identity, provided $\zeta_1, \ldots \zeta_n$ and $\Lambda(A, B)\zeta_1, \ldots \Lambda(A, B)\zeta_n$ lie in $\mathscr{T}_n$. If $\Lambda\zeta_1, \ldots \Lambda\zeta_n$ is outside $\mathscr{T}_n$, the right-hand side is not defined initially, and so (2–85) is no longer an equality between known functions. In that case, we can try to use (2–85) to extend the domain of f_α outside $\mathscr{T}_n$. We get a definition of f_α as a function of n four-vectors $z_1, \ldots z_n$ at all points of the extended tube $\mathscr{T}_n'$; if $z_1, \ldots z_n \in \mathscr{T}_n'$ we define $f(z_1, \ldots z_n)$ by (2–85), where $z_j = \Lambda\zeta_j$, $[\zeta_j \in \mathscr{T}_1, \Lambda \in L_+(C)]$. It is not trivial to show that if a point $z \in \mathscr{T}_n'$ can be reached from two different points of $\mathscr{T}_n$, say ζ and w, by complex Lorentz transformations, then the two ways of defining $f_\alpha(z_1, \ldots z_n)$, via (2–85), lead to the same value. That is, we have yet to show that the continuation is single-valued. Thus, if $\zeta_1, \ldots \zeta_n \in \mathcal{T}_n$, $w_1, \ldots w_n \in \mathcal{T}_n$, and

$$z_j = \Lambda(A_1, B_1)\zeta_j = \Lambda(A_2, B_2)w_j \qquad j = 1, 2, \ldots n$$

we have to prove that

$$\sum_\beta S(A_1, B_1)_{\alpha\beta} f_\beta(\zeta_1, \ldots \zeta_n) = \sum_\beta S(A_2, B_2)_{\alpha\beta} f_\beta(w_1, \ldots w_n),$$

or equivalently, if we use the multiplication law of the group,

$$f_\alpha(\zeta_1, \ldots \zeta_n) = \sum_\beta S(A_1^{-1}A_2, B_1^{-1}B_2)_{\alpha\beta} f_\beta(w_1, \ldots w_n)$$
$$\zeta_1, \ldots \zeta_n \in \mathcal{T}_n, \; w_1, \ldots w_n \in \mathcal{T}_n. \qquad (2\text{–}86)$$

If we put $A = A_1^{-1}A_2$, $B = B_1^{-1}B_2$, then $\zeta_j = \Lambda(A, B)w_j$, and (2–86) reduces to (2–85). Thus the single valuedness is ensured if (2–85) holds for all $\Lambda \in L_+(C)$ with $\zeta_1, \ldots \zeta_n \in \mathcal{T}_n$ and $\Lambda\zeta_1, \ldots \Lambda\zeta_n \in \mathcal{T}_n$. So far we have proved (2–85) only if Λ lies in some neighborhood of $L_+^\uparrow$, that is, if Λ is nearly real. The main part of the proof that (2–85) holds for all complex $\Lambda \in L_+(C)$ is the following lemma, illustrated in Figure 2–2.

Lemma

Suppose $\zeta_1, \ldots \zeta_n \in \mathcal{T}_n$ and $\Lambda\zeta_1, \ldots \Lambda\zeta_n \in \mathcal{T}_n$ with $\Lambda \in L_+(C)$. Then there exists a continuous curve of proper complex

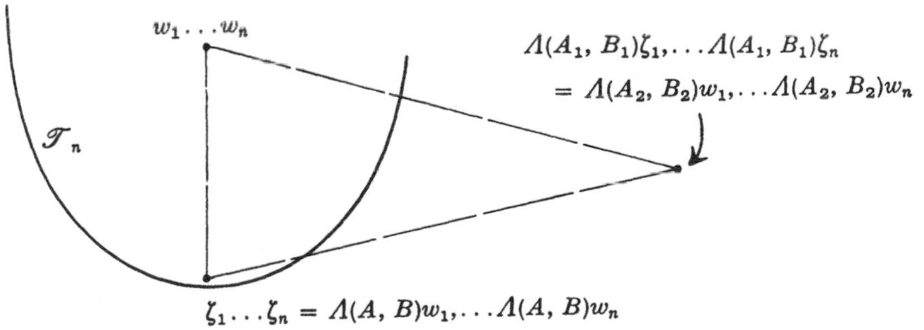

FIGURE 2–2. The single-valuedness of the functions as defined throughout the extended tube $\mathcal{T}_n'$ follows from the fact that, if $w_1, \ldots w_n$ and $\Lambda w_1, \ldots \Lambda w_n \in \mathcal{T}_n$, then there is a curve of proper complex Lorentz transformations $\{\Lambda(t); 0 \leqslant t \leqslant 1\}$ with $\Lambda(0) = 1$, $\Lambda(1) = \Lambda$, such that $\Lambda(t)w_1 \ldots \Lambda(t)w_n \in \mathcal{T}_n$ for $0 \leqslant t \leqslant 1$. Here $A = A_1^{-1}A_2$, $B = B_1^{-1}B_2$.

Lorentz transformations $\{\Lambda(t), 0 \leqslant t \leqslant 1\}$ such that $\Lambda(0) = 1$, $\Lambda(1) = \Lambda$ and $\Lambda(t)\zeta_1, \ldots \Lambda(t)\zeta_n \in \mathscr{T}_n$ for all $0 \leqslant t \leqslant 1$.

We shall prove this lemma in a minute; using it we can prove

Theorem 2–11

If $f_\alpha(\zeta_1, \ldots \zeta_n)$ transforms according to (2–84) under $SL(2,C)$, and is holomorphic in the tube $\eta_j \in V_+$, where $\zeta_j = \xi_j - i\eta_j$, $j = 1, 2, \ldots n$, then $f_\alpha(\zeta_1, \ldots \zeta_n)$ possesses a single-valued analytic continuation into the extended tube $\mathscr{T}'_n$, and transforms according to (2–85) under $L_+(C)$.

Proof:

Suppose $w_j = \Lambda\zeta_j$, $j = 1, 2, \ldots n$, where $\Lambda \in L_+(C)$ and $w_1, \ldots w_n \in \mathscr{T}_n$, $\zeta_1, \ldots \zeta_n \in \mathscr{T}_n$, and let $\Lambda(t)$ be the curve in $L_+(C)$ given by the lemma. We know that (2–85) holds if Λ is in some complex neighborhood of the identity, and this includes a portion of the curve $\Lambda(t)$, say $0 \leqslant t \leqslant t_1$. Since $\Lambda(t_1)\zeta_1, \ldots \Lambda(t_1)\zeta_n \in \mathscr{T}_n$ we can apply (2–85) to some complex neighborhood of $\Lambda(t_1)$, and using the group property of $S(A, B)$ we show (2–85) holds for all $\Lambda(t)$ on the curve up to, say, $t_2 > t_1$. Proceeding in this way (the well-known method of analytic continuation by means of overlapping neighborhoods) in a finite number of steps we see that (2–85) holds for $\Lambda = \Lambda(1)$. The remarks made above then show that the continuation into $\mathscr{T}'$ given by (2–85) is single-valued. That the function so defined is holomorphic is an immediate consequence of the identities

$$\frac{\partial}{\partial[\Lambda(A, B)\zeta_j]^\mu} f_\alpha(\Lambda(A, B)\zeta_1, \ldots \Lambda(A, B)\zeta_n) = \sum_\beta S(A, B)_{\alpha\beta} \sum_{\nu=0}^{3} \frac{\partial f_\beta(\zeta_1, \ldots \zeta_n)}{\partial \zeta_j^\nu} \frac{\partial \zeta_j^\nu}{\partial[\Lambda(A, B)\zeta_j]^\mu}. \quad \blacksquare$$

Proof of the Lemma:

It is convenient to work with the 2×2 matrix formalism described in Section 1–3, in which $\Lambda(A, B)$ is realized by the transformation

$$\underset{\sim}{z} \to A\underset{\sim}{z}B^T \qquad \text{(we write } \underset{\sim}{z} \text{ instead of } \underset{\sim}{\zeta}\text{).}$$

We can use the invariance of the tube $\mathscr{T}_n$ under restricted real Lorentz transformations to bring the A and B into a simple form so that the construction of the required curve is especially easy. We have

$$CA\underset{\sim}{z}B^TC^* = [CA(\bar{B})^{-1}C^{-1}]C\bar{B}\underset{\sim}{z}B^TC^*.$$

For any B, C of determinant 1, $\underset{\sim}{z} \to C\bar{B}\underset{\sim}{z}B^TC^*$ is a restricted real Lorentz transformation. Furthermore, by a suitable choice of the matrix C, the square bracket can be brought into one of the two Jordan canonical forms†

$$K = \pm\begin{pmatrix} 1 & 1 \\ 0 & 1 \end{pmatrix} \tag{2-87}$$

or

$$K = \begin{Bmatrix} \exp(\alpha + i\beta) & 0 \\ 0 & \exp -(\alpha + i\beta) \end{Bmatrix} \qquad \alpha, \beta \text{ real.} \tag{2-88}$$

Thus, it suffices to show that the curves can be constructed for the Lorentz transformations, $\underset{\sim}{z} \to K\underset{\sim}{z}$. We can immediately drop the case of the minus sign in (2–87) because *that* K carries all vectors in the tube out of the tube. That leaves two cases for which we propose the curves

$$K(t) = \begin{Bmatrix} 1 & t \\ 0 & 1 \end{Bmatrix} \qquad 0 \leqslant t \leqslant 1, \tag{2-89}$$

and

$$K(t) = \begin{Bmatrix} \exp[t(\alpha + i\beta)] & 0 \\ 0 & \exp[-t(\alpha + i\beta)] \end{Bmatrix} \qquad 0 \leqslant t \leqslant 1, \tag{2-90}$$

respectively.

Consider first (2–89). If we write $\underset{\sim}{z}(t) = K(t)\underset{\sim}{z}$ we have

$$\underset{\sim}{z}(0) = \underset{\sim}{z}, \qquad \underset{\sim}{z}(1) = \underset{\sim}{z} + \begin{pmatrix} 0 & 1 \\ 0 & 0 \end{pmatrix}\underset{\sim}{z}$$

and therefore

$$\underset{\sim}{z}(t) = \underset{\sim}{z} + t\begin{pmatrix} 0 & 1 \\ 0 & 0 \end{pmatrix}\underset{\sim}{z} = (1 - t)\underset{\sim}{z}(0) + t\underset{\sim}{z}(1) \qquad 0 \leqslant t \leqslant 1.$$

This shows that $\underset{\sim}{z}(t)$ is a convex linear combination of $\underset{\sim}{z}(0)$ and $\underset{\sim}{z}(1)$. But $\mathscr{T}_1$ is convex, so if $\underset{\sim}{z}(0) \in \mathscr{T}_1$ and $\underset{\sim}{z}(1) \in \mathscr{T}_1$, $(1 - t)\underset{\sim}{z}(0) + t\underset{\sim}{z}(1) \in \mathscr{T}_1$. Therefore the curve $K(t)\zeta_1, \ldots K(t)\zeta_n$ lies in $\mathscr{T}_n$ if its end points $\zeta_1, \ldots \zeta_n$ and $K\zeta_1, \ldots K\zeta_n$ do.

For the other case, we shall also exploit the convexity of the tube but the required argument is a bit more arduous. We shall use the following criterion for a real vector to be in the open cone V_+: $y \in V_+$ if and only if $n \cdot y > 0$ for every $n \in C_+$, $n \neq 0$, where C_+ is the bound-

† See P. R. Halmos, *Finite-Dimensional Vector Spaces*, 2nd Ed., Van Nostrand, Princeton, N.J., 1958, p. 113.

ary of V_+. This can be proved with a straightforward application of Schwarz's inequality for the space parts of the vectors. It is convenient to use the notation

$$y^a = y^0 + y^3 \qquad y^b = y^0 - y^3$$

so the scalar product may be written

$$n \cdot y = \tfrac{1}{2}(n^a y^b + n^b y^a) - n^1 y^1 - n^2 y^2.$$

Clearly, for every vector $\in C_+$, $n^a \geqslant 0$, $n^b \geqslant 0$, and $n^a + n^b = 2n^0 > 0$, so at least one of n^a and n^b must be positive.

Let us write $\underset{\sim}{z}(t) = K(t)\underset{\sim}{z}$, $\underset{\sim}{z}(0) = \underset{\sim}{z}$. We make one more real restricted Lorentz transformation,

$$\begin{aligned}\underset{\sim}{\rho}(t) &= K(-t/2)\underset{\sim}{z}(t)K(-t/2)^* \\ &= K(t/2)\underset{\sim}{z}K(-t/2)^*.\end{aligned}$$

Then $z(t) \in \mathscr{T}_1$ if and only if $\rho(t) \in \mathscr{T}_1$. Let $\rho^\mu(t) = \xi^\mu(t) - i\eta^\mu(t)$. The proof of the lemma is complete if we show $\eta(t) \in V_+$, $0 \leqslant t \leqslant 1$. To obtain the vector of the matrix $\underset{\sim}{\rho}(t)$, we evaluate

$$\underset{\sim}{\rho}(t) = \begin{pmatrix} e^{t/2(\alpha + i\beta)} & 0 \\ 0 & e^{-t/2(\alpha + i\beta)} \end{pmatrix}$$

$$\times \begin{pmatrix} z^a & z^1 - iz^2 \\ z^1 + iz^2 & z^b \end{pmatrix} \begin{pmatrix} e^{-t/2(\alpha - i\beta)} & 0 \\ 0 & e^{t/2(\alpha - i\beta)} \end{pmatrix},$$

that is,

$$\begin{pmatrix} \rho^a(t) & \rho^1(t) - i\rho^2(t) \\ \rho^1(t) + i\rho^2(t) & \rho^b(t) \end{pmatrix} = \begin{pmatrix} e^{i\beta t} z^a & e^{\alpha t}(z^1 - iz^2) \\ e^{-\alpha t}(z^1 + iz^2) & e^{-i\beta t} z^b \end{pmatrix}$$

Thus

$$\eta^a(t) = -\frac{1}{2i}(e^{i\beta t} z^a - e^{-i\beta t}\overline{z^a}) = y^a \cos \beta t - x^a \sin \beta t$$

where $z^\mu = x^\mu - iy^\mu$. Similarly, $\eta^b(t) = y^b \cos \beta t + x^b \sin \beta t$,

$$\begin{aligned}\eta^1(t) &= -\frac{\mathrm{Im}}{2}[e^{\alpha t}(z^1 - iz^2) + e^{-\alpha t}(z^1 + iz^2)] \\ &= y^1 \cosh \alpha t + x^2 \sinh \alpha t \\ \eta^2(t) &= -\frac{\mathrm{Re}}{2}[e^{\alpha t}(z^1 - iz^2) - e^{-\alpha t}(z^1 + iz^2)] \\ &= -x^1 \sinh \alpha t + y^2 \cosh \alpha t.\end{aligned}$$

The first step of the proof is to show that $\eta^a(t) > 0$ and $\eta^b(t) > 0$ for $0 \leqslant t \leqslant 1$.

We remark that since the sign of $\sin\beta$ can be changed by replacing x^a, x^b by $-x^a$, $-x^b$, we may assume $0 < \beta < \pi$ without loss in generality; $\beta = 0$ is trivial; $\beta = \pi$ is not possible, because then $\eta^a(1) < 0$, $\eta^b(1) < 0$. For $0 < \beta < \pi$ we have the identity

$$(\sin\beta)\,\eta^a(t) = \sin[(1-t)\beta]\eta^a(0) + (\sin\beta t)\,\eta^a(1)$$

$$(\sin\beta)\,\eta^b(t) = \sin[(1-t)\beta]\eta^b(0) + (\sin\beta t)\,\eta^b(1).$$

This displays $\eta^a(t)$ and $\eta^b(t)$ as positive linear combinations of their initial and final values, and so proves that they are positive.

The second step uses the condition mentioned above for the vector $\eta(t)$ to lie in V_+. Let $n \in C_+$ be a fixed light-like vector and set

$$g(t) = n\cdot\eta(t) = f_1(t) - f_2(t)$$

$$f_1(t) = \tfrac{1}{2}[n^a\eta^b(t) + n^b\eta^a(t)]$$

$$f_2(t) = n^1\eta^1(t) + n^2\eta^2(t),$$

or, substituting,

$$f_1(t) = K_1\cos\beta t + K_2\sin\beta t$$

$$f_2(t) = \lambda_1 e^{\alpha t} + \lambda_2 e^{-\alpha t},$$

where K_1, K_2, λ_1, λ_2 are some real constants. We easily see that

$$\frac{d^2 f_1}{dt^2} = -\beta^2 f_1, \qquad \frac{d^2 f_2}{dt^2} = \alpha^2 f_2,$$

$$\frac{d^2 g}{dt^2} = -\beta^2 f_1 - \alpha^2 f_2 = \alpha^2 g - (\alpha^2 + \beta^2) f_1.$$

Because $\eta(0)$, $\eta(1) \in V_+$ we have $g(0) > 0$, $g(1) > 0$. Of course $g(t)$ is an elementary function, and if $g(t) < 0$ in the interval $(0, 1)$ it must have a minimum in the interval; for the minimum

$$\frac{d^2 g}{dt^2} \geqslant 0, \qquad \text{i.e., } g \geqslant \frac{(\alpha^2 + \beta^2) f_1}{\alpha^2} > 0.$$

Therefore $g(t) > 0$ in $(0, 1)$, which proves the lemma. ∎

Since holomorphic functions with a transformation law such as (2–84) have single-valued analytic continuations to the extended tube, it is of some interest to characterize the extent of this domain more precisely. We will not indulge in this sport here, except to determine its *real* points which are of great importance in the *PCT* theorem.

By its very definition, the tube $\mathscr{T}_n$ does not contain real points: $z_1, \ldots z_n \in \mathscr{T}_n$ requires $-\text{Im}\, z_j \in V_+$, $j = 1, 2, \ldots n$ and therefore not zero. However, the extended tube $\mathscr{T}_n'$ does contain real points, commonly called *Jost points*, as will now be explained.

We first consider the special case of one vector. If $\zeta \in \mathscr{T}_1$, $\zeta = \xi - i\eta$, $\eta \in V_+$, and the extended tube $\mathscr{T}_1'$ consists of all points $\Lambda\zeta$, $\Lambda \in L_+(C)$, $\zeta \in \mathscr{T}_1$. Since $\Lambda\zeta \cdot \Lambda\zeta = \zeta \cdot \zeta$, the values of ζ^2 for ζ in $\mathscr{T}_1'$ are the same as those for ζ in $\mathscr{T}_1$. Now $\zeta^2 = \xi^2 - \eta^2 - 2i\xi \cdot \eta$, so if ζ^2 is real and $\zeta \in \mathscr{T}_1$, ξ is orthogonal to a time-like vector and therefore is space-like. Thus $\zeta^2 < 0$. Hence a real point of $\mathscr{T}_1'$ must have $\zeta^2 < 0$. This condition is also sufficient for $\zeta \in \mathscr{T}_1'$; for if $\zeta^2 < 0$, and ζ is real, we can choose a coordinate system such that $\zeta = (\zeta^0, \zeta^1, 0, 0)$ with $\zeta^1 > |\zeta^0|$; then the complex Lorentz transformation

$$\begin{aligned} \zeta^0 + \zeta^1 &\to e^{i\alpha}(\zeta^0 + \zeta^1) = \tilde{\zeta}^0 + \tilde{\zeta}^1 \\ \zeta^0 - \zeta^1 &\to e^{-i\alpha}(\zeta^0 - \zeta^1) = \tilde{\zeta}^0 - \tilde{\zeta}^1 \end{aligned} \tag{2-91}$$

gives $\tilde{\zeta}^0 = i \sin\alpha\zeta^1 + \cos\alpha\zeta^0$, $\tilde{\zeta}^1 = i \sin\alpha\zeta^0 + \cos\alpha\zeta^1$. This transformation takes ζ into the tube if $\sin\alpha < 0$ since then $\text{Im}\,\tilde{\zeta}^0 < -|\text{Im}\,\tilde{\zeta}^1|$. In the general case, we have the following theorem due to Jost.

A real point $\zeta_1, \ldots \zeta_n$ lies in the extended tube $\mathscr{T}_n'$ if and only if all vectors of the form

$$\sum_{j=1}^{n} \lambda_j \zeta_j, \qquad \lambda_j \geqslant 0, \sum \lambda_j > 0$$

are space-like; that is,

$$\left(\sum_{j=1}^{n} \lambda_j \zeta_j\right)^2 < 0 \qquad \text{for all } \lambda_j \geqslant 0 \text{ with } \sum \lambda_j > 0. \tag{2-92}$$

Proof:

We first prove the necessity of (2–92). By definition $\zeta_1, \ldots \zeta_n \in \mathscr{T}_n'$, if there exists a proper complex Lorentz transformation Λ such that $\Lambda\zeta_1, \ldots \Lambda\zeta_n \in \mathscr{T}_n$. Now $(\Lambda\zeta_j)^2 = \zeta_j^2$, and we have already seen that z^2 real, $z \in \mathscr{T}_1$ is possible only if $z^2 < 0$. Therefore $\zeta_j^2 < 0$, $j = 1, 2, \ldots n$. Moreover, $\mathscr{T}_1$ is convex so that if each of the complex four-vectors $\Lambda\zeta_1, \ldots \Lambda\zeta_n$ is in $\mathscr{T}_1$, every linear combination of the form

$$\sum_{j=1}^{n} \lambda_j \Lambda\zeta_j \qquad \text{for } \lambda_j \geqslant 0 \qquad j = 1, 2, \ldots n, \sum \lambda_j > 0$$

is in $\mathcal{T}_1$. Therefore

$$\left[\Lambda\left(\sum_{j=1}^{n}\lambda_j\zeta_j\right)\right]^2 < 0,$$

which reduces to

$$\left(\sum_j \lambda_j\zeta_j\right)^2 < 0.$$

This proves the necessity of (2–92). To prove the sufficiency, suppose the vectors $\zeta_1, \ldots \zeta_n$ satisfy (2–92). They then span a convex space-like cone K, which intersects neither the plus nor the minus light cone. We can find two planes (α), (β), touching the forward and backward cones, respectively, which separate K from these cones (see Figure 2–3). This is possible because any two such convex sets can be so separated.†

Let the equations of (α), (β) be

$$\alpha_\mu\zeta^\mu = 0, \qquad \beta_\mu\zeta^\mu = 0$$

α and β are light-like, and $\alpha\cdot\beta < 0$. We can choose a coordinate system so that $\alpha = (1, 1, 0, 0)$, $\beta = (-1, +1, 0, 0)$.

For points in V_+, $\zeta\cdot\alpha > 0$, so for points below the plane α, $\zeta\cdot\alpha < 0$, which, in the particular coordinate system, means

$$\zeta^0 - \zeta^1 < 0 \qquad \text{if } \zeta \in K.$$

Similarly, if $-\zeta \in V_+$, $\beta\cdot\zeta > 0$, so $-\zeta^0 - \zeta^1 < 0$ for points above (β), that is, points in K.

Therefore $\zeta^1 > |\zeta^0|$ for all points in K, in particular for $\zeta_1, \ldots \zeta_n$ given. We therefore can apply the complex Lorentz transformation used above, (2–91), which takes ζ_k into the forward tube, $k = 1, 2, \ldots n$. This shows the sufficiency of (2–92). ∎

Theorem 2–11 shows that any functions holomorphic in $\mathcal{T}_n$ and satisfying a transformation law such as (2–84) possess single-valued analytic continuations to $\mathcal{T}_n'$. However, nothing which has been said so far excludes that they can be continued even farther. For

† Instead of using this general property of convex sets one can argue more directly. We know that if ρ is a time-like and ζ a space-like vector then $\alpha\rho + \beta\zeta$ cannot be zero unless both α and β are zero. Now we look for solutions, n, of the sets of simultaneous inequalities $n\cdot\rho < 0$, $n\cdot\zeta < 0$, where ρ runs over the minus light cone, V_-, and ζ over the convex cone spanned by the $\zeta_1, \ldots \zeta_n$. It is easy to see that this set of inequalities will have *no* solution if and only if the convex cone spanned by the ρ's and ζ's contains the zero vector. From the above argument this is impossible, so there is at least one solution, n. Then the plane $n\cdot\zeta = 0$ is a separating plane, and it is not difficult to see that one of the solutions lies in C_+, the boundary of V_+.

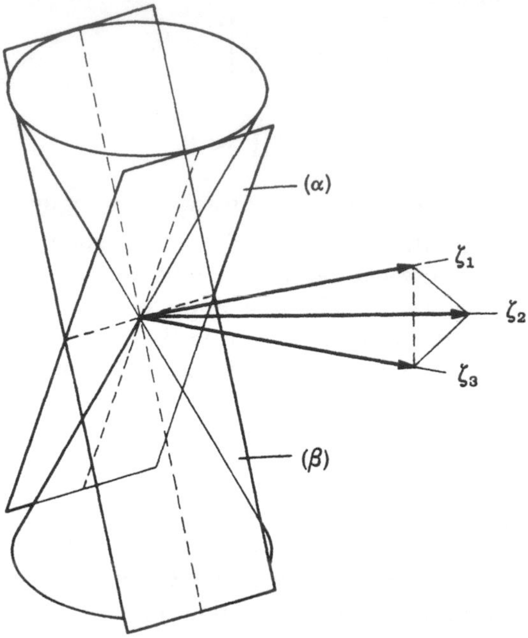

FIGURE 2–3. The cone of space-like vectors corresponding to a Jost point $\zeta_1, \zeta_2, \zeta_3$. The planes (α) and (β) separate the cone spanned by $\zeta_1, \zeta_2, \zeta_3$ from the interior of the plus and minus light cones.

real points Theorem 2–12 leads the way to simple examples which show that such a continuation is generally impossible: the function

$$\int_0^\infty \cdots \int_0^\infty dK\, d\lambda_1 \ldots d\lambda_n \frac{\rho(K, \lambda_1, \ldots \lambda_n)}{\left(\sum_{j=1}^{n} \lambda_j \zeta_j\right)^2 - K^2}$$

is analytic in $\mathscr{T}_n$, invariant under Lorentz transformations but singular at all real points which are not Jost points if ρ is suitably chosen.

It is clear that the Jost points of $\mathscr{T}'_n$ form a real environment for holomorphic functions in $\mathbf{C}^{4n}$; thus if a holomorphic function vanishes at Jost points, it vanishes everywhere.

Later on, in our study of the commutation relations of fields, we shall be concerned not only with the extended tube $\mathscr{T}'_n$ but also with the so-called *permuted extended tubes*. There is one of these domains for each element of the permutation group, S_{n+1}, of $n + 1$ objects, and they are obtained by linear transformations from $\mathscr{T}'_n$. For later purposes it suffices to discuss the particular domain which corresponds to the transposition of the j with the $j+1$ object. The corresponding linear transformation is denoted $P(j, j+1)$ and is defined by

$$\begin{aligned} \hat{\zeta}_k &= \zeta_k \qquad 1 \leqslant k < j-1 \quad \text{and} \quad j+1 < k \leqslant n \\ \hat{\zeta}_{j-1} &= \zeta_{j-1} + \zeta_j \\ \hat{\zeta}_j &= -\zeta_j \\ \hat{\zeta}_{j+1} &= \zeta_{j+1} + \zeta_j, \end{aligned} \tag{2–93}$$

with the convention that if a ζ appears with index <1 or $>n$ the equation is to be deleted. All the rest of the permuted extended tubes can be generated from these.

Our purpose here is to establish that $\mathscr{T}'_n$ and $P(j, j+1)\mathscr{T}'_n$ have a real environment in common. To this end, we choose $\zeta_k = (0, b, 0, 0)$ for $k \neq j-1$, j, or $j+1$, and $\zeta_{j-1} = (a, b, 0, 0)$, $\zeta_j = (0, 0, \varepsilon, 0)$, $\zeta_{j+1} = (-a, b, 0, 0)$, where $0 < |a| < b$. All the vectors, $\lambda_{j-1}\zeta_{j-1} + \lambda_j\zeta_j + \lambda_{j+1}\zeta_{j+1}$ with λ's $\geqslant 0$ and not all zero, are space-like, and so are all the vectors $\lambda_{j-1}(\zeta_{j-1} + \zeta_j) - \lambda_j\zeta_j + \lambda_{j+1}(\zeta_{j+1} + \zeta_j)$ (see Figure 2–4). It is easy to see that adjoining the rest of the ζ_k gives a Jost point of both $\mathscr{T}'_n$ and $P(j, j+1)\mathscr{T}'_n$, and that the same is true for all real points in a sufficiently small neighborhood of these.

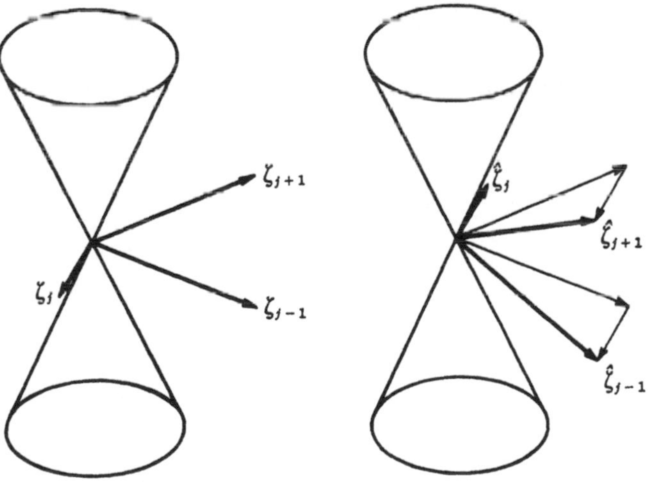

FIGURE 2.4 A configuration of vectors ζ_{j-1}, ζ_j, ζ_{j+1} appearing for a Jost point of $\mathscr{T}'_n$ and $P(j, j+1)\ \mathscr{T}'_n$. The $\hat{\zeta}$'s are obtained from the ζ's by $\hat{\zeta}_{j-1} = \zeta_{j-1} + \zeta_j$, $\hat{\zeta}_j = -\zeta_j$, $\hat{\zeta}_{j+1} = \zeta_{j+1} + \zeta_j$.

2–5. THE EDGE OF THE WEDGE THEOREM

In its simplest form for one complex variable the edge of the wedge theorem is ancient and well known. We prove it in the form given by Painlevé in 1888.

Theorem 2–13

Let F_1 be a function holomorphic in an open set D_1 in the upper half-plane with an open interval $a < x < b$ as part of its boundary. Let F_2 be holomorphic in an open set D_2 in the lower half-plane with $a < x < b$ as part of its boundary. Suppose

$$F_1(x) = \lim_{y \to 0+} F_1(x + iy) \tag{2–94}$$

and

$$F_2(x) = \lim_{y \to 0+} F_2(x - iy) \tag{2–95}$$

exist uniformly, in $a < x < b$, are continuous and satisfy

$$F_1(x) = F_2(x) \qquad \text{for } a < x < b.$$

Then F_1 and F_2 are actually holomorphic on $a < x < b$ and are the same holomorphic function.

Remark:

The assumption of uniformity in the theorem is, in fact, redundant. It can be shown that if the limit functions F_1 and F_2 exist and are continuous then the convergence is uniform. However, to make the discussion completely elementary, we make the hypothesis of uniformity.

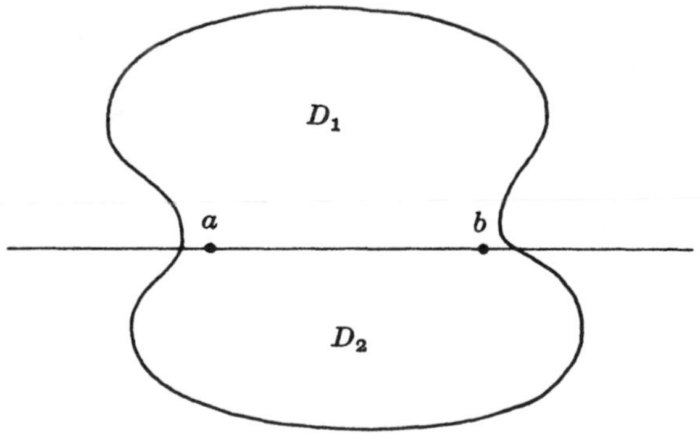

FIGURE 2–5. Domains D_1 and D_2, of definition of F_1 and F_2, respectively.

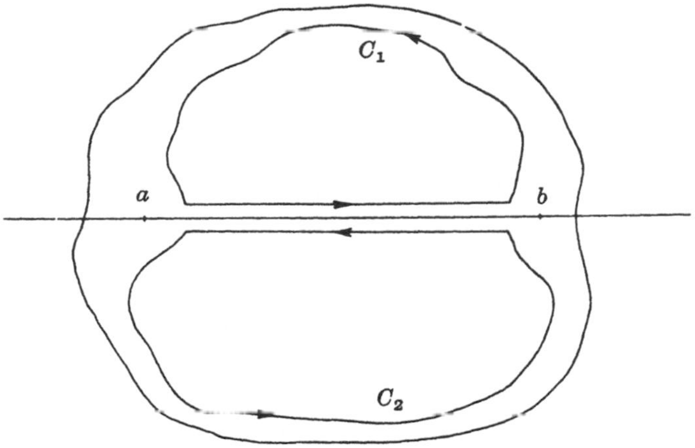

FIGURE 2–6. The contours C_1 and C_2.

Proof:

Let C_1 and C_2 be two contours lying in D_1 and D_2 respectively except for the interval $a' \leqslant x \leqslant b'$, where $a < a' < b' < b$. Then

$$\frac{1}{2\pi i}\int_{C_1}\frac{F_1(\xi)\,d\xi}{(\xi - z)} = \begin{cases} F_1(z) & z \text{ inside } C_1 \\ 0 & z \text{ inside } C_2 \end{cases}$$

and

$$\frac{1}{2\pi i}\int_{C_2}\frac{F_2(\xi)\,d\xi}{(\xi - z)} = \begin{cases} F_2(z) & z \text{ inside } C_2 \\ 0 & z \text{ inside } C_1 \end{cases}$$

by Cauchy's integral formula. Further, because the limit is uniform,

$$\lim_{\varepsilon\to 0}\int_{a'}^{b'}\frac{F_j(\xi \pm i\varepsilon)}{\xi - z \pm i\varepsilon}\,d\xi = \int_{a'}^{b'}\frac{F_j(\xi)}{\xi - z}\,d\xi \qquad j = 1, 2, \text{ respectively.}$$

Now examine the function G defined by

$$G(z) = \frac{1}{2\pi i}\left\{\int_{C_1}\frac{F_1(\xi)\,d\xi}{(\xi - z)} + \int_{C_2}\frac{F_2(\xi)\,d\xi}{(\xi - z)}\right\} = \frac{1}{2\pi i}\int_C\frac{F(\xi)\,d\xi}{(\xi - z)},$$

where C is the contour obtained by following $C_1 - (a'b')$ and then $C_2 - (b'a')$. F is defined as F_1 above and F_2 below the real axis. G is holomorphic throughout the interior of C_1 and C_2 and also on the open interval $a' < x < b'$. It coincides with F_1 inside C_1 and with

F_2 inside C_2. Thus it provides the required analytic continuation of F_1 and F_2. ∎

A corollary of this theorem is the familiar Schwarz reflection principle, which says that if F_1 is holomorphic in D_1 as in the theorem and converges uniformly to boundary values on every subinterval of $a < x < b$ which define a real continuous function, then F_1 is holomorphic in $\bar{D}_1$ and $\overline{F_1(\bar{z})}$ continues it into $\bar{D}_1$. Here $\bar{D}_1$ is the domain complex conjugate to D_1. The Schwarz reflection principle actually antedated Theorem 2–13 and inspired it.

We shall generalize this theorem in stages, eventually arriving at the edge of the wedge theorem in the form required in later chapters.

The first step is to pass from one to several complex variables. That essentially new features appear is already visible for two variables. The analogue of the region D_1 lies in the product of the upper half-planes: $y_1 > 0$ and $y_2 > 0$; the analogue of D_2 lies in the product of the lower half-planes: $y_1 < 0$ and $y_2 < 0$. The assertion of the theorem is holomorphy in some neighborhood of the interval on the real axis where F_1 and F_2 coincide. That neighborhood will necessarily include a chunk of the regions: $y_1 > 0$ and $y_2 < 0$, and $y_1 < 0$ and $y_2 > 0$. This is a set of new points of holomorphy of the same dimension as that of the space, and the assertion is in that respect a stronger statement than Theorem 2–13. (It is related to the phenomenon of *analytic completion*; see Bochner and Martin, Chapter IV.) The size of the region of new points of holomorphy depends on the size of the regions D_1 and D_2. This is an inevitable feature of the situation which somewhat complicates the statement of the theorem.

Theorem 2–14

Let $\mathcal{O}$ be an open set of $\mathbf{C}^n$ which contains a real environment E which is an open set of $\mathbf{R}^n$. Suppose D_1 is the intersection of $\mathcal{O}$ with the product of the upper half-planes:

$$D_1 = \{z_1, \ldots z_n;\ z_1, \ldots z_n \in \mathcal{O},\ y_1 > 0, \ldots y_n > 0\} \quad (2\text{–}96)$$

and D_2 is the intersection of $\mathcal{O}$ with the product of the lower half-planes:

$$D_2 = \{z_1, \ldots z_n;\ z_1, \ldots z_n \in \mathcal{O},\ y_1 < 0, \ldots y_n < 0\}. \quad (2\text{–}97)$$

Assume F_1 holomorphic in D_1, F_2 in D_2, and the limits, for $y_1 > 0, \ldots y_n > 0$,

$$\lim_{y_1,\ldots y_n \to 0} F_1(x_1 + iy_1, \ldots x_n + iy_n) = F_1(x_1, \ldots x_n) \quad (2\text{–}98)$$

$$\lim_{y_1,\ldots y_n \to 0} F_2(x_1 - iy_1, \ldots x_n - iy_n) = F_2(x_1, \ldots x_n) \quad (2\text{–}99)$$

exist, are continuous in $x_1, \ldots x_n$ and are equal for $x_1 \ldots x_n \in E$, the limit being uniform in E.

Then there exists a (complex) neighborhood, N, of E in $\mathbf{C}^n$, and a holomorphic function, G, such that G coincides with F_1 on D_1 and F_2 on D_2, and is holomorphic on N. The neighborhood N can be chosen to depend only on $\mathcal{O}$ and E.

Remarks:

1. G evidently provides an analytic continuation of F_1 and F_2.

2. Since E is an open set it is a union of cubes. Thus, it suffices to prove the theorem for E an open cube. Furthermore, since by real translation and a real change of scale such a cube can be centered about the origin and adjusted to any size, without loss of generality, we may assume that E is of a convenient size, say $-1 - \varepsilon < x_j < 1 + \varepsilon$, $\varepsilon > 0$, $j = 1, \ldots n$.

3. In the course of the proof we are going to use a contour of integration in the z_j-plane which is a circle through the points -1 and $+1$. The contour is satisfactory provided it lies in $\mathcal{O}$.

4. The actual proof we shall carry out uses something stronger than is stated in the hypothesis of the theorem: the boundedness of the derivatives $\partial F_1/\partial z_j$, $\partial F_2/\partial z_j$ in the neighborhood of the points $z_1 \ldots z_n = +1, \ldots +1$ and $-1, \ldots -1$. Given functions F_1, F_2 with the property in the theorem we can always define *primitives*

$$\hat{F}_j(z_1, \ldots z_n) = \int_{-1}^{z_1} d\zeta_1 \ldots \int_{-1}^{z_n} d\zeta_n\, F_j(\zeta_1, \ldots \zeta_n) \qquad j = 1, 2$$

and, because of the uniform convergence of F,

$$\lim_{y_1,\ldots y_n \to 0} \hat{F}_1(z_1, \ldots z_n) = \int_{-1}^{x_1} d\xi_1 \ldots \int_{-1}^{x_n} d\xi_n\, F_1(\xi_1, \ldots \xi_n)$$
$$= \lim_{y_1,\ldots y_n \to 0} \hat{F}_2(\bar{z}_1, \ldots \bar{z}_n).$$

The first derivatives of $\hat{F}$ are bounded. Clearly the primitives are continuous on $\bar{D}_1$, $\bar{D}_2$ and are holomorphic at the same points as F_1, F_2. Thus there is no loss in generality in assuming $\partial F_1/\partial z_j$, $\partial F_2/\partial z_j$ are bounded.

Proof:

Define the function G of $n + 1$ complex variables in the domain $|\zeta| < 1$, $|z_j| < R$, $j = 1, \ldots n$ by the formula

$$G(\zeta, z_1, \ldots z_n) = \frac{1}{2\pi i} \int_{C_+} \frac{F_1\left(\dfrac{u + z_1}{1 + uz_1}, \ldots \dfrac{u + z_n}{1 + uz_n}\right) du}{u - \zeta}$$

$$+ \frac{1}{2\pi i} \int_{C_-} \frac{F_2\left(\dfrac{u + z_1}{1 + uz_1}, \ldots \dfrac{u + z_n}{1 + uz_n}\right) du}{u - \zeta}, \tag{2–100}$$

where the contours C_+ and C_- in the u plane are the unit semi-circles traversed counterclockwise:

$$\begin{aligned} C_+ &: \quad |u| = 1 \qquad \operatorname{Im} u \geqslant 0, \\ C_- &: \quad |u| = 1 \qquad \operatorname{Im} u \leqslant 0. \end{aligned} \tag{2–101}$$

If R is sufficiently small and the cube has been chosen sufficiently small, the arguments of the functions F_1 and F_2 will be entirely in D_1 and D_2, respectively, as u varies over C_+ and C_-, respectively, by virtue of the elementary identity

$$\operatorname{Im}\left(\frac{u + z}{1 + uz}\right) = \frac{\eta(1 - |z|^2) + y(1 - |u|^2)}{|1 + uz|^2},$$

where $u = \xi + i\eta$, $z = x + iy$ (see Figure 2–7). The preceding statement holds except when $u = \pm 1$, in which case the arguments are $\pm 1, \ldots \pm 1$, both points of E.

It is clear from the continuity of F_1 and F_2 that (2–100) defines G as a continuous function of all its variables together, which is analytic in ζ for fixed values of $z_1, \ldots z_n$. Furthermore, the partial derivatives of the G with respect to the z_j exist by our assumption that the partial derivatives of F_1 and F_2 are bounded in the neighborhood of E. Therefore, G is a holomorphic function of $\zeta, z_1, \ldots z_n$ in $|\zeta| < 1$, $|z_j| < R$, $j = 1, \ldots n$. In particular, its values $G(0, z_1, \ldots z_n)$, as a function of $z_1, \ldots z_n$ for $\zeta = 0$, define a holomorphic function of n variables in $|z_j| < R$, $j = 1, \ldots n$. It remains to identify $G(0, z_1, \ldots z_n)$ with $F_1(z_1, \ldots, z_n)$ in D_1 and F_2 in D_2.

Consider $G(\zeta, x_1, \ldots x_n)$ with $|x_j| < R$, $j = 1, \ldots n$. Then the integrands in the first and second terms are respectively

$$F_j\left(\frac{u + x_1}{1 + ux_1}, \ldots \frac{u + x_n}{1 + ux_n}\right) \qquad j = 1, 2. \tag{2–102}$$

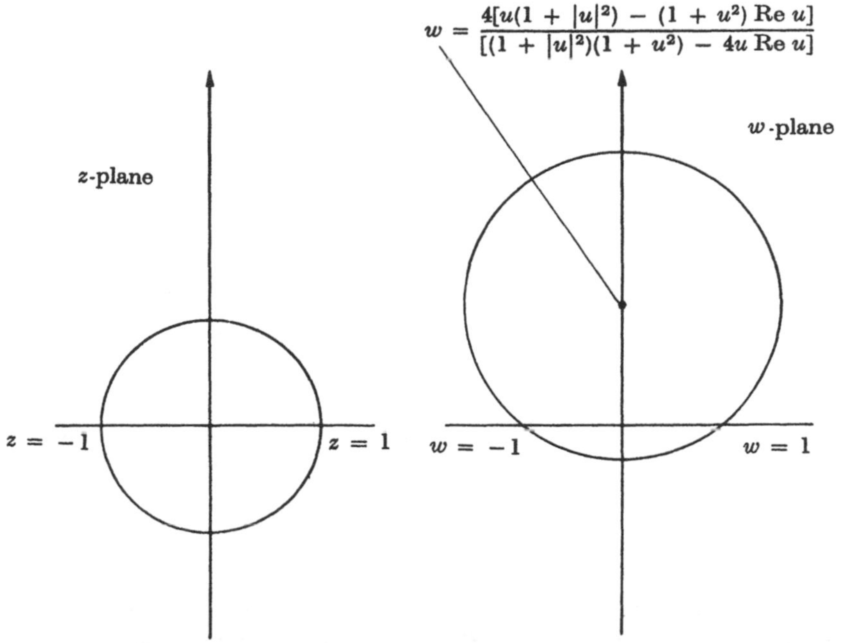

FIGURE 2–7. Illustration of the mapping $w = (u + z)/(1 + uz)$. As z runs over the unit circle $|z| = 1$, w runs over the circle indicated. The points $z = \pm 1$ always are mapped into $w = \pm 1$.

They are holomorphic as functions of u for $\mathrm{Im}\, u > 0$ and $\mathrm{Im}\, u < 0$, respectively, and have equal boundary values as $\mathrm{Im}\, u = 0$ by assumption.† By Theorem 2–13, they are therefore analytic throughout the unit disc and continue one another. Thus, in this case, (2–100) is Cauchy's formula:

$$G(\zeta, x_1, \ldots x_n) = \begin{cases} F_1\left(\dfrac{\zeta + x_1}{1 + \zeta x_1}, \dfrac{\zeta + x_2}{1 + \zeta x_2}, \ldots \dfrac{\zeta + x_n}{1 + \zeta x_n}\right) & \mathrm{Im}\,\zeta > 0 \\ F_2\left(\dfrac{\zeta + x_1}{1 + \zeta x_1}, \dfrac{\zeta + x_2}{1 + \zeta x_2}, \ldots \dfrac{\zeta + x_n}{1 + \zeta x_n}\right) & \mathrm{Im}\,\zeta < 0. \end{cases} \tag{2–103}$$

Now for fixed ζ the left-hand side is the restriction to a real environment of the holomorphic function $G(\zeta, \ldots)$, while depending on the sign of the imaginary part of ζ the right-hand side is the restriction of

† By the above remark, the F_j are continuous in all the variables together.

$$F_1\left(\frac{\zeta + z_1}{1 + \zeta z_1}, \ldots \frac{\zeta + z_n}{1 + \zeta z_n}\right),$$

or

$$F_2\left(\frac{\zeta + z_1}{1 + \zeta z_1}, \ldots \frac{\zeta + z_n}{1 + \zeta z_n}\right).$$

Thus $G(\zeta, \ldots)$ provides an analytic continuation of each of these. In particular, for $\zeta = 0$, $G(0, z_1, \ldots z_n)$ coincides with $F_1(z_1, \ldots z_n)$ in D_1 and $F_2(z_1, \ldots z_n)$ in D_2 and is itself analytic in $|z_j| < R, j = 1, \ldots n$. Clearly, by construction R has been chosen to depend only on D_1, D_2, and E. ∎

Next, we want to show how Theorem 2–14 can be used to get an edge of the wedge theorem for a region of apparently much more general shape.

Theorem 2–15

Let $\mathcal{O}$ be an open set of $\mathbf{C}^n$ which contains a real environment, E, with E some open set of $\mathbf{R}^n$. Let $\mathscr{C}$ be an open convex cone of $\mathbf{R}^n$. Suppose F_1 is holomorphic in

$$D_1 = (\mathbf{R}^n + i\mathscr{C}) \cap \mathcal{O} \qquad (2\text{–}104)$$

and F_2 in

$$D_2 = (\mathbf{R}^n - i\mathscr{C}) \cap \mathcal{O}. \qquad (2\text{–}105)$$

(The notation $\mathbf{R}^n \pm i\mathscr{C}$ stands for the sets of all vectors of the form $x \pm iy$, respectively, where x and y are real vectors of n components and $y \in \mathscr{C}$.)

Suppose the limits for $x \in E$

$$\lim_{\substack{y \to 0 \\ y \in \mathscr{C}}} F_1(x + iy) = F_1(x) \qquad (2\text{–}106)$$

and

$$\lim_{\substack{y \to 0 \\ y \in \mathscr{C}}} F_2(x - iy) = F_2(x) \qquad (2\text{–}107)$$

exist and are continuous and equal on E, the limit being uniform on E.

Then there is a (complex) neighborhood, N, of E, and a holomorphic function, G, which coincides with F_1 in D_1 and F_2 in D_2 and is holomorphic in N. Here N is independent of F_1 and F_2 but of course depends on E, $\mathscr{C}$, and $\mathcal{O}$.

Proof:

Since $\mathscr{C}$ is an open convex cone we can choose n linearly independent vectors in it, say $y^{(k)}$, $k = 1, \ldots n$. The convex cone they span is a subcone $\mathscr{C}'$ of $\mathscr{C}$. Since the $y^{(k)}$ are linearly independent, any complex n vector z may be written

$$z = \sum_{k=1}^{n} \lambda_k y^{(k)}.$$

The variables $\lambda_1, \ldots \lambda_n$ are related to the $z_1, \ldots z_n$ by a non-singular real linear transformation. When z runs over $\mathbf{R}^n + i\mathscr{C}'$, the λ_k vary independently in the upper half-plane. Thus, the hypotheses of the present theorem (when $\mathscr{C}$ is replaced by $\mathscr{C}'$ and the z's are replaced by λ's) reduce to those of Theorem 2–14. Its conclusions re-expressed in z's are those above. ∎

In the applications of Chapter 4, we shall deal with sets of n four-vectors $\zeta_1, \ldots \zeta_n$, where $\zeta_j = \xi_j - i\eta_j$ and $\eta_j \in V_+$. Thus $\mathscr{C}$ in that case is a cone in $4n$ dimensions, the direct product of n plus light cones.

Now we introduce a second essential complication in the edge of the wedge theorem. We admit that F_1 and F_2 need not have boundary values which are continuous functions approached pointwise, but rather may have boundary values which are distributions approached in the sense of distribution theory. A familiar example is the function $(x + iy)^{-1}$ for $y \to 0$. Its boundary value as $y \to 0$, $y > 0$, is $P\left(\frac{1}{x}\right) - \pi i \, \delta(x)$, in the sense that for each test function f

$$\lim_{y \to 0} \int \frac{f(x)\, dx}{x + iy} = P \int \frac{f(x)\, dx}{x} - \pi i f(0),$$

where P stands for the Cauchy principal value. That is just what we have defined as convergence of distributions in $\mathscr{D}'$. Thus the refined version of the edge of the wedge theorem is

Theorem 2–16

Suppose that the hypotheses of Theorem 2–15 hold except that instead of (2–106) and (2–107) we have for every test function of compact support lying in E

$$\lim_{\substack{y \to 0 \\ y \in \mathscr{C}}} \int F_1(x + iy)\, dx\, f(x) = T(f)$$

and

$$\lim_{\substack{y \to 0 \\ y \in \mathscr{C}}} \int F_2(x - iy)\, dx\, f(x) = T(f),$$

where T is a distribution of $\mathscr{D}(E)'$.

The conclusions of Theorem 2–15 again hold.

Proof:

The idea of the proof is as follows. Regularize F_1 and F_2 to get F_{1f} and F_{2f}:

$$\begin{aligned} F_{1f}(x + iy) &= \int d\xi\, f(x - \xi) F_1(\xi + iy) \\ F_{2f}(x + iy) &= \int d\xi\, f(x - \xi) F_2(\xi + iy), \end{aligned} \tag{2–108}$$

where f is an infinitely differentiable function of sufficiently small support. F_{1f} and F_{2f} are again holomorphic functions but now satisfy the assumptions of Theorem 2–15. The resulting G_f then has to be shown to be of the form $G_f(x + iy) = \int f(x - \xi)\, d\xi\, G(\xi + iy)$, where G is the required holomorphic function.

Equation (2–108) indeed defines two holomorphic functions when the support of f is sufficiently small and $x + iy$ is confined to some subset of D_1 or D_2 respectively, because the integrals are continuous in $x + iy$ and satisfy the Cauchy–Riemann equations in each variable separately;

$$\begin{aligned} \frac{\partial}{\partial x_j} \int d\xi\, f(x - \xi) F(\xi + iy) &= \int d\xi\, \frac{\partial f}{\partial x_j}(x - \xi) F(\xi + iy) \\ &= \int d\xi\, f(x - \xi) \frac{\partial F}{\partial \xi_j}(\xi + iy) \\ &= -i \frac{\partial}{\partial y_j} \int d\xi\, f(x - \xi) F(\xi + iy). \end{aligned} \tag{2–109}$$

As $y \to 0$, these holomorphic functions converge to $T(\hat{f}_{-x})$ uniformly for x on suitably small compact subsets of E. Here $\hat{f}_{-x}(\xi) = f(x - \xi)$. (Recall that the convergence of a sequence of distributions is always uniform on bounded sets if it converges at all, and the $\hat{f}_x$ form a bounded set as x varies over a compact set.) Thus, by Theorem 2–15 there exists a holomorphic function G_f which coincides with F_{1f} in an open subset of D_1 and with F_{2f} in an open subset of D_2, is holomorphic in a neighborhood of an open subset of E, and provides an analytic continuation of F_1 and F_2. The construction of G_f displays it as a Cauchy integral with integrand which is a distribution in f, so $G_f(x + iy)$ is, for fixed

$x + iy$, a distribution in f. Thus, by the Schwartz nuclear theorem, there is a distribution $H(\xi, x + iy)$ such that

$$G_f(x + iy) = \int d\xi\, f(-\xi)H(\xi, x + iy).$$

Now an examination of the Cauchy integral formula for G_f yields immediately that for sufficiently small t

$$G_{f_t}(x + iy) = G_f(x + t + iy)$$

which implies, because of the uniqueness of H,

$$H(\xi + t, x + iy) = H(\xi, x + t + iy).$$

In other words, H depends only on $x + \xi$, and not on $\xi - x$; that is,

$$G_f(x + iy) = \int d\xi\, f(x - \xi)H(\xi, iy).$$

Now let f_k be a sequence of test functions which converge to a δ-function at the origin and g be a test function in x and y. Then, for any test function g of sufficiently small support,

$$\int dx\, dy\, g(x, y)G_{f_k}(x + iy) = \int d\xi\, dy\, [\int dx\, f_k(x - \xi)g(x, y)]H(\xi, iy)$$
$$\to \int dx\, dy\, g(x, y)H(x, iy).$$

But the $G_{f_k}(x + iy)$ is a sequence of holomorphic functions, and, as we have already explained in Section 2–3, the convergence for all g of such a sequence smeared with test functions of support lying in or on a given compact set implies the uniform convergence of the sequence to a holomorphic function, $G(x + iy)$. Therefore $H(x, iy) = G(x + iy)$ is a holomorphic function. Since the sequence converges on an open set of D_1 to F_1 and on an open set of D_2 to F_2, it provides the required analytic continuation. ∎

In the applications of Chapter 4 we shall have occasion to use the following theorem, which is an easy consequence of the edge of the wedge theorem.

Theorem 2–17

Let $\mathcal{O}$ be an open set of $\mathbf{C}^n$, containing a real environment E which is an open set of $\mathbf{R}^n$. Suppose F is a function holomorphic in

$$\mathcal{B} = (\mathbf{R}^n + i\mathcal{C}) \cap \mathcal{O}$$

where $\mathcal{C}$ is an open convex cone of $\mathbf{R}^n$. Suppose further that

$$\lim_{y \to 0} F(x + iy) = 0 \qquad \text{for } x \in E, \qquad (2\text{–}110)$$

where the convergence is understood in the sense of $\mathscr{D}'(E)$.
Then $F = 0$ throughout $\mathscr{B}$.

Proof:

Define $F_1(x + iy) = \overline{F(x - iy)}$. The function F_1 is holomorphic in the domain $\overline{\mathscr{B}}$, complex conjugate to $\mathscr{B}$, and approaches 0 as $y \to 0$. Thus we can apply the edge of the wedge theorem (2–16), to F and F_1 and conclude that there is a G holomorphic in a complex neighborhood of E which is an analytic continuation of F and F_1. But according to (2–110), G vanishes on E as a distribution and therefore as a function. Since E is a real environment, G vanishes throughout $\mathscr{B}$. ∎

2–6. HILBERT SPACE

Throughout Chapter 1, we have taken for granted that the reader is acquainted with the elementary theory of Hilbert space. Here we make a series of disconnected remarks which we hope will relate the elementary theory to some of the arguments used elsewhere in the book.

Recall that a Hilbert space, $\mathscr{H}$, is a vector space with scalars which are complex numbers,† and a scalar product (,) which is a complex-valued function of two vectors in $\mathscr{H}$ satisfying

$$(\Phi, \Psi) = \overline{(\Psi, \Phi)}, \qquad (\Phi, \alpha\Psi + \beta\chi) = \alpha(\Phi, \Psi) + \beta(\Phi, \chi),$$

and

$$\|\Phi\|^2 \equiv (\Phi, \Phi) \geqslant 0, \tag{2–111}$$

and

$$(\Phi, \Phi) = 0 \qquad \text{implies } \Phi = 0. \tag{2–112}$$

Furthermore, $\mathscr{H}$ is required to be *complete* in the sense that any Cauchy sequence of vectors has a limit. That means if Φ_n, $n = 1, 2, \ldots$ is a sequence of vectors such that for each $\varepsilon > 0$ there is an integer N such that

$$\|\Phi_n - \Phi_m\| < \varepsilon \qquad \text{for all } n, m \geqslant N$$

(then Φ_n, $n = 1, 2, \ldots$ is called a *Cauchy sequence*), there exists a vector Φ such that

$$\lim_{n\to\infty} \|\Phi_n - \Phi\| = 0.$$

† There are of course also real Hilbert spaces which use real scalars but they will not be considered here.

The first point we want to discuss is the separability of the Hilbert spaces occurring in quantum field theory. Recall that a set S of vectors is *dense* in $\mathscr{H}$ if, for each vector $\Phi \in \mathscr{H}$ and $\varepsilon > 0$, there exists a vector $\Psi \in S$ such that $\|\Phi - \Psi\| < \varepsilon$. A Hilbert space is *separable* if it contains a denumerable dense set, or, in other words, there is a sequence of vectors which is dense. In non-separable Hilbert spaces a continuous index is necessary to label a dense set. An alternative way of describing this distinction is in terms of complete orthonormal sets. A Hilbert space is separable if it contains a denumerable complete orthonormal set; it is non-separable if complete orthonormal sets are not denumerable. One can pass from one description to the other by orthonormalizing the dense set to get a denumerable complete orthonormal set, or forming finite linear combinations of the denumerable complete orthonormal set with complex numbers whose real and imaginary parts are rational to get a denumerable dense set. In von Neumann's original axiomatization the requirement of separability was included as a defining property of Hilbert space. Nowadays, it has become conventional to use the name to cover the non-separable case also. Of late, physicists have started considering vector spaces with a scalar product in which (2–111) and (2–112) are abandoned (indefinite metric). We shall not call such spaces Hilbert spaces, nor, in fact, consider them at all.

In non-relativistic quantum mechanics it is natural to consider only separable Hilbert spaces because one is usually dealing with a finite number of particles and can realize the states as vectors in $L^2(\mathbf{R}^n)$, i.e., as equivalence classes of square integrable functions on n-dimensional euclidean space, two functions being equivalent if they differ on a set of measure zero. It is well known that this Hilbert space is separable (Ref. 20). It is sometimes argued that in quantum field theory one is dealing with a system of an infinite number of degrees of freedom and so must use a non-separable Hilbert space. Roughly, the idea is

$$\begin{pmatrix}\text{system of a finite number}\\ \text{of degrees of freedom}\end{pmatrix} \leftrightarrow \begin{pmatrix}\text{separable Hilbert}\\ \text{space}\end{pmatrix}$$

$$\begin{pmatrix}\text{system of an infinite number of}\\ \text{degrees of freedom or something}\end{pmatrix} \leftrightarrow \begin{pmatrix}\text{non-separable Hilbert}\\ \text{space}\end{pmatrix}$$

Our next task is to explain why this is wrong, or at best grossly misleading.

In the first place note that if $\mathscr{H}_1, \mathscr{H}_2, \mathscr{H}_3, \ldots$ is a sequence of separable Hilbert spaces then their *direct sum* $\bigoplus_j \mathscr{H}_j$ is a separable Hilbert space. The direct sum has elements which are sequences $\{\Phi_1, \Phi_2, \ldots\}$ with $\Phi_j \in \mathscr{H}_j$ and

$$\sum_{j=1}^{\infty} \|\Phi_j\|^2 < \infty.$$

The scalar product in $\bigoplus_j \mathscr{H}_j$ is

$$(\Phi, \Psi) = \sum_{j=1}^{\infty} (\Phi_n, \Psi_n)$$

where (Φ_n, Ψ_n) is the scalar product in $\mathscr{H}_n$ and Φ stands for $\{\Phi_1, \Phi_2, \ldots\}$, Ψ for $\{\Psi_1, \Psi_2, \ldots\}$. It is not hard to see that sequences which have only a finite number of elements different from zero, and the non-zero ones belonging to denumerable dense sets of their respective $\mathscr{H}_j$ form a denumerable dense set in $\bigoplus_j \mathscr{H}_j$.

The direct sum can be used to construct state vectors which are superpositions of states for an arbitrarily large number of particles. One merely takes $\bigoplus_n \mathscr{H}_n$, where $n = 0, 1, 2, \ldots$ and $\mathscr{H}_n$ describes the states of n particles. Just this Hilbert space will be used in Chapter 3 to give explicit formulas for the theory of a free field. This is a clear-cut counter example to the above assertion; a free field is a system of an infinite number of degrees of freedom. Of course, this example does not describe interacting particles, but, to the extent that the arguments of Chapter 1 are correct, the presence of interaction makes no difference. The collision states of an asymptotically complete theory span a separable Hilbert space, because they are again just of the form $\bigoplus_n \mathscr{H}_n$. [Here, $\mathscr{H}_n$ is the subspace spanned by the ingoing (or outgoing) collision states with n colliding particles.] All these arguments make it clear that there is no evidence that separable Hilbert spaces are not the natural state spaces for quantum field theory.

When, in fact, do non-separable Hilbert spaces appear in quantum mechanics? There are two cases which deserve mention. The first arises when one takes an infinite tensor product of separable Hilbert spaces. We shall not give the rather technical definition of infinite tensor product here but only remark that it is a natural generalization of the ordinary tensor product which is used to describe a composite system. Infinite tensor products of Hilbert spaces (of dimension greater than 1!) are always non-separable. Since a (Bose) field can be thought as a system composed of an infinity of oscillators, one might think that such an infinite tensor product is the natural state space.

However, it is characteristic of field theory that some of its observables involve all the oscillators at once and it turns out that such observables can be naturally defined only on vectors belonging to a tiny separable subset of the infinite tensor product. It is the subspace spanned by such a subset which is the natural state space rather than the whole infinite tensor product itself. Thus, while it may be a matter of convenience to regard the state space as part of the infinite tensor product, it is not necessary.

A second example of the occurrence of non-separable Hilbert spaces appears in statistical mechanics when one passes to the limit in which the conventional box containing the system becomes arbitrarily large, the density being maintained constant. Two states of the limiting system which have different densities actually differ by the presence of an infinite number of particles. One might expect them to be orthogonal and in fact that is the case in all examples worked out so far. Thus there is an orthonormal system labeled by a continuous parameter, the density, and the Hilbert space is non-separable. It seems to us that such phenomena are consequences of considering systems in which every state contains an actual infinity of physical particles and provide no argument for analogous phenomena in relativistic quantum field theory. This completes our discussion of separability.

Now we turn to some remarks on linear operators in Hilbert space. Later on in the book there is a good deal of muttering about domains of unbounded operators. There is a general feeling among physicists that anything that depends in an important way on such matters cannot be physics. We would like to offer some arguments to the contrary.

Recall that an *operator*, A, from a Hilbert space $\mathscr{H}_1$ to a Hilbert space $\mathscr{H}_2$ is a function defined on a subset of $\mathscr{H}_1$, called the *domain* of A, $D(A)$, and taking values in $\mathscr{H}_2$, the set of those values being called the *range* of A, $R(A)$. The *graph*, $\Gamma(A)$, of A is the set of all pairs $\{\Phi, A\Phi\}$ with $\Phi \in D(A)$. It is a subset of $\mathscr{H}_1 \oplus \mathscr{H}_2$. A is *linear* if its graph is a linear manifold in $\mathscr{H}_1 \oplus \mathscr{H}_2$. This amounts to saying that if Φ_1 and $\Phi_2 \in D(A)$, then $\alpha\Phi_1 + \beta\Phi_2 \in D(A)$ and

$$A(\alpha\Phi_1 + \beta\Phi_2) = \alpha A\Phi_1 + \beta A\Phi_2,$$

which is the usual definition of linearity. A is *closed* if its graph is a closed set in $\mathscr{H}_1 \oplus \mathscr{H}_2$, that is, if Φ_n, $n = 1, 2, \ldots \in D(A)$, and $\lim_{n\to\infty} \|\Phi_n - \Phi\| = 0$, and $\lim_{n\to\infty} \|A\Phi_n - \Psi\| = 0$, then $\Phi \in D(A)$ and $A\Phi = \Psi$. An operator B is an *extension* of A if $\Gamma(A) \subset \Gamma(B)$. That means $D(A) \subset D(B)$ and $A\Phi = B\Phi$ for all $\Phi \in D(A)$. We write $A \subset B$ in this case.

If S is any set in $\mathscr{H}_1 \oplus \mathscr{H}_2$ its *orthogonal complement* is the set $S^\perp$ of all vectors orthogonal to every vector S. $S^\perp$ is always a closed linear manifold. If S happens to be the graph of an operator from $\mathscr{H}_1$ to $\mathscr{H}_2$, it may happen that $S^\perp$ is the graph of an operator from $\mathscr{H}_2$ to $\mathscr{H}_1$. Whether it is or not depends entirely on whether $\{\Phi, \Psi\}$ and $\{\chi, \Psi\} \in S^\perp$ implies $\Phi = \chi$; if not, the presumed operator would not be single-valued, as it must be by the definition of the notion operator. The condition that $\{\Phi, \Psi\}$ and $\{\chi, \Psi\} \in S^\perp$ is

$$(\{\Phi, \Psi\}, \{\Phi_1, A\Phi_1\}) = (\Phi, \Phi_1) + (\Psi, A\Phi_1) = 0$$

and

$$(\{\chi, \Psi\}, \{\Phi_1, A\Phi_1\}) = (\chi, \Phi_1) + (\Psi, A\Phi_1) = 0 \qquad \text{for all } \Phi_1 \in D(A).$$

Subtracting we see that $(\Phi - \chi, \Phi_1) = 0$, so that if $D(A)$ is dense, $\Phi = \chi$ and $S^\perp$ is the graph of a transformation. The negative of this transformation is called the *adjoint* A^* of A. In other words, A^* is the transformation from $\mathscr{H}_2$ to $\mathscr{H}_1$ defined for just those $\Psi \in \mathscr{H}_2$ such that there exists a $\chi \in \mathscr{H}_1$ satisfying

$$(\Psi, A\Phi) = (\chi, \Phi) \qquad \text{for all } \Phi \in D(A).$$

Then, by definition, $\chi = A^*\Psi$. Clearly, by definition A^* is a closed linear transformation when it exists at all.

The adjoint plays an important role in the construction of closed linear extensions of a given operator. Clearly, any closed linear extension of A must be an extension of the operator whose graph is $\overline{\Gamma(A)}$, the closure in $\mathscr{H}_1 \oplus \mathscr{H}_2$ of $\Gamma(A)$. Since $\overline{\Gamma(A)} = \{[\Gamma(A)]^\perp\}^\perp$, the preceding criterion says that $(A^*)^*$ is this minimal closed linear extension of A if the domains of A and A^* are dense so that both A^* and $(A^*)^*$ exist.

An operator from $\mathscr{H}_1$ to $\mathscr{H}_1$ is *hermitian* if $A \subset A^*$, and *self-adjoint* if $A = A^*$. (Sometimes these are respectively denoted symmetric and hypermaximal-symmetric.) The operator is *essentially self-adjoint* if $A^* = (A^*)^*$. Clearly, A is hermitian if and only if

$$(\Phi, A\Psi) = (A\Phi, \Psi) \qquad \text{for all } \Phi, \Psi \in D(A), \text{ and } D(A) \text{ is dense.}$$

This is what is customarily meant by hermitian in elementary quantum mechanics, but what is relevant if A is to be observable is self-adjointness. Only then are the eigenfunctions of A complete. (For a full explanation see Ref. 20.) Evidently, essential self-adjointness is just as good as self-adjointness from this point of view because $(A^*)^*$ is uniquely determined by A.

An operator, A, is *bounded* if $\|A\Phi\|/\|\Phi\|$ is bounded as Φ runs over $D(A)$, that is,

$$\|A\Phi\| \leqslant M\|\Phi\| \qquad \Phi \in D(A)$$

for some M independent of Φ. A bounded linear operator is evidently continuous because $\|A\Phi - A\Psi\| \leqslant M\|\Phi - \Psi\|$. Conversely, a linear operator is continuous at every point of its domain if it is continuous at 0, and if it is continuous at 0 it is bounded. The first of these statements is an obvious consequence of linearity. The second requires only the following simple argument. Suppose $\|A\Phi\| < \varepsilon$ for all $\Phi \in D(A)$ such that $\|\Phi\| < \delta$. Then, if Ψ is any element of $D(A)$,

$$\Psi\left(\frac{\delta}{2\|\Psi\|}\right) \in D(A) \quad \text{and} \quad \left\|\Psi\left(\frac{\delta}{2\|\Psi\|}\right)\right\| < \delta \qquad \text{so} \left\|A\left(\Psi\left(\frac{\delta}{2\|\Psi\|}\right)\right)\right\| < \varepsilon,$$

that is, $\|A\Psi\| < (2\varepsilon/\delta)\|\Psi\|$, proving A is bounded. A bounded linear operator A always has a linear extension whose domain is all of $\mathscr{H}_1$. We shall outline the proof.

One first extends A by continuity to all limit points of $D(A)$, defining A at the limit point as the limit of its values at neighboring points of $D(A)$. This extension is just $(A^*)^*$. If the closure of $D(A)$ is all of $\mathscr{H}_1$, the construction is complete. If not, one can define A arbitrarily (but linearly and boundedly) on the orthogonal complement of $D(A)$ and complete the definition to $\mathscr{H}_1$ by linearity. If A is an operator from $\mathscr{H}_1$ to $\mathscr{H}_1$ and is hermitian, linear, and bounded, clearly it can always be arranged that its extension to all of $\mathscr{H}_1$ is self-adjoint. Thus in a consideration of bounded linear operators, there is no loss in generality in regarding them as everywhere defined.

The situation for unbounded linear operators is completely different. An unbounded linear operator is discontinuous at every point of its domain. Furthermore, if it is closed it cannot be everywhere defined because a closed everywhere-defined linear operator is necessarily bounded. (This last statement is the *closed graph theorem*. For a proof see Ref. 21.) We are particularly interested in closed operators for two reasons. On the one hand, we need to study self-adjoint extensions of hermitian operators A. An hermitian A always has a closed linear extension $(A^*)^*$ and any self-adjoint extension of A is an extension of $(A^*)^*$. On the other hand, we shall want to study non-hermitian operators A with dense domains, and their adjoints are always closed. It is clear that the best we can hope for in these situations are operators which are densely but not everywhere defined and everywhere discontinuous on their domains.

These facts are part of what is generally regarded as the pathology of Hilbert space, and a natural reaction to them is to make a declaration that only bounded operators will be considered: the only operators are bounded operators! There are even good physical grounds for this position, for, as is explained at great length in Ref. 20, an unbounded operator which represents an observable must be self-adjoint and statements about observations of it can be re-expressed in terms of its spectral projections which are bounded operators. There are several reasons why, nevertheless, the axioms of Chapter 3 have been expressed in terms of unbounded operators. First, it is these quantities, corresponding directly to the classical fields, which are the source of inspiration for the quantum theory of fields. Second, the equations which describe local interaction between fields in space-time are expressed in terms of such unbounded operators. It may be said that these arguments express exactly what is wrong with quantum field theory and that is a defensible view, but the main point of the enterprise which Chapters 3 and 4 represent is the exploration, with all the resources of modern mathematics, of the physical ideas which have developed in the course of the last fifty years. To him who wants to make a more radical alteration in the foundations we say Bravo! and Bon voyage!

Having once resigned ourselves to dealing with unbounded operators, we face a couple of practical problems having to do with their domains. It can happen that A^2 is undefined except on the zero vector because it may be that

$$\{AD(A)\} \cap D(A) = \{0\}.$$

What is worse, we shall have to deal with sets of unbounded operators and shall want to compute polynomials in them. It requires a special assumption to be sure that such operations make sense even on a dense set of vectors. One is led naturally to assume a common dense domain, D, for all the unbounded operators in question on which all polynomials are applicable. Once one has D, another problem arises: to what extent do the values of the unbounded operators on D determine them uniquely wherever else they can be defined? This is particularly acute for observables because an operator which is hermitian, when restricted to vectors in D, might have several different self-adjoint extensions, and to specify a theory one would have to tell which self-adjoint extension is meant. Furthermore, equations valid on vectors of D need not be valid for other vectors. For example, one can have two operators, each essentially self-adjoint when restricted to vectors on a dense domain D and commuting on such vectors, but whose self-adjoint extensions do not commute (Ref. 23). Whether such behavior can happen in relativistic quantum field theory is not known at the moment, and it is somewhat amazing how complete a theory one can get without settling it. The main point of the preceding remarks is to bring these important problems to the reader's attention. The rest of the book is devoted to by-passing them.

The remaining subject we will discuss is the theory of continuous unitary representations of the translation group of space-time. Here the main point is to connect the statements of Chapter 1 about the energy momentum spectrum of a physical theory to the criteria which will be used in Chapters 3 and 4. More specifically, what we have in mind is this. In Chapters 3 and 4 we will have matrix elements of the form

$$(\Phi, U(a, 1)\Psi) \tag{2–113}$$

for certain special vectors Φ and Ψ, and will argue that

$$\int e^{-ip\cdot a}\, da(\Phi, U(a, 1)\Psi) = 0 \tag{2–114}$$

unless p belongs to the physical spectrum which lies in $\bar{V}_+$. The operation indicated in (2–114) is mathematically meaningful because (2–113) will be shown to be an infinitely differentiable bounded function, and so has a Fourier transform as a tempered distribution. It makes sense to assert that a tempered distribution vanishes in an open set. On the other hand, how do we know that this mathematical assertion is related to the statement that the theory has an energy momentum spectrum in $\bar{V}_+$? Formally, this is usually stated by writing an expansion over intermediate states of physical momentum Q and some other quantum numbers α:

$$(\Phi, \Psi) = \sum_\alpha \int dQ \langle \Phi \mid Q\alpha\rangle\langle Q\alpha \mid \Psi\rangle.$$

Then

$$\int da\, e^{-ip\cdot a}(\Phi, U(a, 1)\Psi) = \sum_\alpha \int dQ \int da\, e^{-i(p-Q)\cdot a}\langle \Phi \mid Q\alpha\rangle\langle Q\alpha \mid \Psi\rangle$$

$$= (2\pi)^4 \sum_\alpha \int dQ\, \delta(Q - p)\langle \Phi \mid Q\alpha\rangle\langle Q\alpha \mid \Psi\rangle$$

because

$$\langle Q\alpha \mid U(a, 1) \mid \Psi\rangle = e^{iQ\cdot a}\langle Q\alpha \mid \Psi\rangle.$$

These manipulations are not rigorous, because the states $\mid Q, \alpha\rangle$ may have infinite norm. The method can be completely justified by using the theory of direct integrals (Ref. 22), but we want to avoid that.

Alternatively, we can look a little more closely at the form of $U(a, 1)$. According to the SNAG theorem (Ref. 22), $U(a, 1)$ may be written as a Stieltjes integral over momentum space

$$U(a, 1) = \int e^{ip\cdot a}\, dE(p),$$

where E is a projection valued measure on momentum space. That means that for each sphere of momentum space, and for each set S which can be obtained from spheres by a denumerable number of operations of taking unions, intersections, and passing to complements, a projection operator $E(S)$ is defined. $E(S)$ has to satisfy

$$E(S_1)E(S_2) = E(S_1 \cap S_2),$$

$$\sum_{j=1}^{\infty} E(S_j) = E(\bigcup_j S_j) \qquad \text{if } S_j \cap S_k \text{ is empty} \qquad \text{for } j \neq k$$

and

$$E(\mathbf{R}^4) = 1.$$

In terms of E, the statement that a set S is not in the physical energy momentum spectrum is simply $E(S) = 0$ or, equivalently,

$$\int da\, \rho(a)U(a, 1) = \int \tilde{\rho}(p)\, dE(p) = 0$$

for all $\rho \in \mathscr{S}$ such that supp $\tilde{\rho}$ lies in S. Here

$$\rho(x) = \frac{1}{(2\pi)^2}\int e^{-ip\cdot a}\tilde{\rho}(p)\, dp.$$

Evidently, this implies

$$\int da\, \rho(a)(\Phi, U(a, 1)\Psi) = 0,$$

which is another way of stating (2–114).

These few words are evidently no substitute for a complete exposition of the SNAG theorem, but they give the essential ideas which make possible a precise statement of the physical hypothesis about the energy momentum spectrum.

The $E(S)$ that can appear in a theory which is invariant under the restricted Lorentz group as well as the translation group cannot be arbitrary. In particular, there is only one p at which a discrete point spectrum is possible: $p = 0$. It is easy to see why this must be. If $P^\mu\Psi_p = p^\mu\Psi_p$ and $(\Psi_p, \Psi_p) = 1$ for some $p \neq 0$, then $U(0, \Lambda)\Psi_p$ would satisfy $P^\mu U(0, \Lambda)\Psi_p = (\Lambda p)^\mu U(0, \Lambda)\Psi_p$, and, as Λ runs over $L_+^\uparrow$, the $U(0, \Lambda)\Psi_p$ would be a continuous family of normalized states orthogonal when $\Lambda_1 p \neq \Lambda_2 p$. That is impossible in a separable Hilbert space. It is interesting that this makes it possible to characterize the vacuum state in simpler ways: it is the only normalizable eigenfunction of P^0, or $\mathbf{P}$, or P^1.

BIBLIOGRAPHY

The standard work on distribution theory is

1. L. Schwartz, *Théorie des distributions*, Hermann, Paris, Part I, 1957, Part II, 1959.

Another systematic treatise is

2. I. Gelfand and co-authors, *Generalized Functions* I...V (in Russian) Gosizdfizmatlit, Moscow, 1958.

A very useful summary for physicists is

3. L. Gårding and J. Lions, "Functional Analysis," *Nuovo Cimento Suppl.*, **14**, 9 (1959). The brief exposition of Sections 2–1 and 2–2 is not a substitute for Gårding–Lions, which in turn, as those authors themselves remark, is not a substitute for Schwartz's book. The proofs of Theorems 2–8, 2–9, and 2–10 follow unpublished remarks of L. Gårding.

The most elementary proof of the nuclear theorem is due to Gelfand and Vilenkin, Ref. 2, Vol. IV. Other proofs may be found in

4. L. Ehrenpreis, "On the Theory of the Kernels of Schwartz," *Proc. Amer. Math. Soc.*, **7**, 713 (1956).
5. H. Gask, "A Proof of Schwartz's Kernel Theorem," *Math. Scand.*, **8**, 327 (1960).

For the Laplace transform, see

6. L. Schwartz, "Transformation de Laplace des distributions," *Medd. Lunds Mat. Sem. Supplementband*, p. 196 (1952).

Holomorphic functions of several complex variables are treated in

7. S. Bochner and W. T. Martin, *Several Complex Variables*, Princeton University Press, Princeton, N.J., 1948.

A summary for physicists, with further references, is

8. A. S. Wightman, "Analytic Functions of Several Complex Variables," pp. 159–221 in *Dispersion Relations and Elementary Particles*, Wiley, New York, 1960.

Theorem 2–11 was first proved for invariant holomorphic functions in the thesis of D. Hall; see

9. D. Hall and A. S. Wightman, "A Theorem on Invariant Analytic Functions with Applications to Relativistic Quantum Field Theory," *Dan. Mat. Fys. Medd.*, **31**, No. 5 (1957).

The extension to sets of fields with an arbitrary transformation law is in

10. R. Jost, "Properties of Wightman Functions" in *Lectures on Field Theory and the Many-Body Problem*, E. R. Caianiello (ed.), Academic Press, New York, 1961; and
11. A. S. Wightman, "Quantum Field Theory and Analytic Functions of Several Complex Variables, *J. Indian Math. Soc.*, **24**, 625 (1960).

The proof of the crucial lemma given in the text is due to V. Bargmann (unpublished).

Theorem 2–12 is due to R. Jost:

12. R. Jost, "Eine Bemerkung zum CTP Theorem," *Helv. Phys. Acta*, **30**, 409 (1957).

Painlevé's proof of Theorem 2–13 appears in

13. P. Painlevé, "Sur les lignes singulières des fonctions analytiques," *Ann. Fac. Toulouse*, **2**, B27 (1888).

The edge of the wedge theorem so named first appears in

14. H. J. Bremmermann, R. Oehme, and J. G. Taylor, "A Proof of Dispersion Relations in Quantized Field Theories," *Phys. Rev.*, **109**, 2178 (1958). The proof given is only valid under very strong restrictions on the behavior of the functions and their derivatives near the boundary, and is not valid for distribution boundary values. Earlier, in the course of a proof of dispersion relations, Bogoliubov, Medvedev, and Polivanov had given arguments which are quite adequate to handle the case of distribution boundary values, but since the whole matter appeared in a specific context they did not try to isolate the edge of the wedge theorem.
15. N. N. Bogoliubov and D. V. Shirkov, *Quantum Theory of Fields*, Interscience, New York, 1958, Russian edition, 1956.
16. N. N. Bogoliubov, B. V. Medvedev, M. K. Polivanov, *Questions of the Theory of Dispersion Relations* (in Russian), Moscow, 1958.

More specifically, the theorem as here stated is a local theorem about holomorphic functions, whereas in the context of dispersion relations one has analytic functions which are typically Laplace transforms of functions with restricted support. See, however, V. S. Vladimirov, *On Bogoliubov's "Edge of the Wedge Theorem,"* *Izvestia Akad. Nauk USSR*, **26**, 825 (1962).

Dyson outlined a proof "without pretensions to rigor" which is very simple and natural. Its ideas are at the heart of the proof we present here.

17. F. J. Dyson, "Connection between Local Commutativity and Regularity of Wightman Functions," *Phys. Rev.*, **110**, 579 (1958). This proof was made precise by Gårding and Beurling in a much quoted article which has never appeared; Theorem 2–16 follows their method.

The edge of the wedge theorem was generalized to the case in which the cones to which the imaginary parts are confined are not opposite by Epstein:

18. H. Epstein, "Generalization of the 'Edge-of-the-Wedge' Theorem," *J. Math. Phys.*, **1**, 524 (1960).

A proof of the edge of the wedge theorem which may have advantages to the mathematician, because of the extensive exercise in functional analysis it provides, is

19. F. Browder, "On the 'Edge of the Wedge' Theorem," *Can. Math. J.*, **15**, 125 (1963).

In the elementary proof of Theorems 2–14 and 2–15 given here we have incorporated unpublished suggestions of H. Borchers and V. Glaser.

The classic introduction to the mathematical foundations of quantum mechanics is

20. J. von Neumann, *Mathematical Foundations of Quantum Mechanics*, Princeton University Press, Princeton, N.J., 1955. The separability of the Hilbert space $L^2(\mathbf{R}^n)$ is established on p. 64.

An economical account of the closed graph theorem is on pp. 17–18 of

21. L. H. Loomis, *Abstract Harmonic Analysis*, Van Nostrand, Princeton, N.J., 1953.

The S, N, A, G in the SNAG theorem stand for Stone, Naimark, Ambrose, and Godement. An account of it may be had by combining Chapter 10 of Ref. 25 with Chapter II of

22. J. Dixmier, *Les algèbres d'opérateurs dans l'espace hilbertien (Algèbres de von Neumann)*, Gauthier Villars, Paris, 1956.

An example of two noncommuting self-adjoint operators which commute on a common dense domain of essential self-adjointness is given in

23. E. Nelson, "Analytic Vectors," *Ann. Math.*, **70**, 572 (1959), especially pp. 603–604.

The separability of $\mathscr{S}$ is proved on p. 373 of

24. G. Köthe, *Topologische lineare Räume*, I, Springer, Berlin, 1960.

Another useful reference on the commutability of unbounded operators is

25. F. Riesz and B. Sz.-Nagy, *Functional Analysis*, Ungar, New York, 1955, especially Section 116.

CHAPTER 3

FIELDS AND VACUUM EXPECTATION VALUES

You boil it in sawdust: You salt it in glue
You condense it with locusts and tape
Still keeping one principal object in view—
To preserve its symmetrical shape.

Fit the Fifth, *The Hunting of the Snark*
LEWIS CARROLL

The classical notion of field originated in attempts to avoid the idea of action at a distance in the description of electromagnetic and gravitational phenomena. In these important cases, the field turns out to have two basic properties: (1) it is observable, and (2) it is defined by a set of functions on space-time with a well-defined transformation law under the appropriate relativity group. Since in quantum mechanics observables are represented by hermitian operators which act on the Hilbert space of state vectors, one expects the analogue in relativistic quantum mechanics of a classical observable field to be a set of hermitian operators defined for each point of space-time and having a well-defined transformation law under the appropriate group. The first part of the present chapter is devoted to the isolation of a mathematical definition of field in quantum mechanics which accords with these general ideas. It turns out that not only observable but also unobservable fields are of interest, so we shall also consider sets of non-hermitian operators. The second part of the chapter shows how a theory of fields can be expressed in terms of certain associated distributions, the vacuum expectation values of products of field operators. The technique of expressing properties of the theory in terms of properties of the vacuum expectation values is the principal tool used in getting the results in Chapter 4.

3–1. AXIOMS FOR THE NOTIONS OF FIELD AND FIELD THEORY

It was recognized early in the analysis of field measurements for the electromagnetic field in quantum electrodynamics that, in their dependence on a space-time point, the components of fields are in general more singular than ordinary functions. This suggests that only smeared fields be required to yield well-defined operators. For example, in the case of the electric field, $\mathscr{E}(\mathbf{x}, t)$ is not a well-defined operator, while $\int d\mathbf{x}\, dt\, f(x)\mathscr{E}(\mathbf{x}, t) = \mathscr{E}(f)$ is. Here f is any infinitely differentiable function of compact support defined in space-time. There is another point that should be mentioned. A smeared field like $\mathscr{E}(f)$ can have an arbitrarily large expectation value in a suitably chosen state; hence we expect to have to deal with unbounded operators. It is known that in general unbounded operators cannot be defined on every vector in any natural way. We shall therefore be obliged to make some assumptions about the domain of vectors on which the smeared fields are definable. Typical of the states on which fields cannot be defined are those for which the expectation value is infinite. This is familiar from elementary quantum mechanics where the position operator is not definable on states $\Psi(\mathbf{x})$ which are normalizable, $\int |\Psi(\mathbf{x})|^2\, d\mathbf{x} < \infty$, but which are such that $\int |\mathbf{x}|^2\, |\Psi(\mathbf{x})|^2\, d\mathbf{x}$ does not converge.

The assumptions we make fall into four groups.

O. Assumptions of Relativistic Quantum Theory

The states of the theory are described by unit rays in a separable Hilbert space $\mathscr{H}$. The relativistic transformation law of the states is given by a continuous unitary representation of the inhomogeneous $SL(2,C)$:

$$\{a, A\} \rightarrow U(a, A).$$

Since $U(a, 1)$ is unitary it can be written as $U(a, 1) = \exp(iP^\mu a_\mu)$, where P^μ is an unbounded hermitian operator, interpreted as the energy momentum operator of the theory. The operator $P^\mu P_\mu = m^2$ is interpreted as the square of the mass. The eigenvalues of P^μ lie in or on the plus cone (*spectral condition*).

There is an invariant state, Ψ_0,

$$U(a, A)\Psi_0 = \Psi_0 \qquad (3\text{–}1)$$

unique up to a constant phase factor (uniqueness of the vacuum).

According to the discussion of Section 2–1 we could, on firm physical grounds, assume much more about the multiplicity of states but that is unnecessary in what follows.

Next we give the properties defining a field, whose transformation law under $SL(2,C)$ is given by the $n \times n$ matrix representation, $S: A \to S(A)$.

I. Assumptions about the Domain and Continuity of the Field

For each test function $f \in \mathscr{S}$, defined on space-time, there exists a set $\varphi_1(f), \ldots \varphi_n(f)$ of operators. These operators, together with their adjoints $\varphi_1(f)^*, \ldots \varphi_n(f)^*$ are defined on a domain D of vectors, dense† in $\mathscr{H}$. Furthermore, D is a linear set containing Ψ_0

$$\Psi_0 \in D \tag{3–2}$$

and the $U(a, A)$ and the $\varphi_j(f)$ and $\varphi_j(f)^*$ carry vectors in D into vectors in D

$$U(a, A)D \subset D, \qquad \varphi_j(f)D \subset D, \qquad \varphi_j(f)^*D \subset D, \tag{3–3}$$

where $j = 1, 2, \ldots n$.

If Φ, $\Psi \in D$, then $(\Phi, \varphi_j(f)\Psi)$ is a tempered distribution, regarded as a functional of f.

Two remarks concerning the domain D are in order. First, D always contains the domain, D_0, consisting of those vectors which are obtained from the vacuum state by applying polynomials in the smeared fields; that is an immediate consequence of (3–2) and (3–3). Second, an unbounded operator, defined and hermitian on a domain, may have several self-adjoint extensions to vectors not in the domain, even if the domain is dense. This is a well-known mathematical phenomenon first analyzed by von Neumann (see Section 2–6 and Ref. 20 of the Bibliography of Chapter 2). It is only for a *self-adjoint* operator that one can assert the usual spectral resolution, completeness of the expansion into eigenfunctions, etc., which are essential features for an operator which describes an observable in quantum mechanics. Thus, for observable fields to be completely specified from a physical point of view, the domain D must be large enough so that such fields when defined on D have unique self-adjoint extensions. It is an attractive conjecture that D_0 is already large enough for this purpose, but that is unproved at the moment. Thus, to make a definition of the notion of field sufficiently flexible to cope with whatever happens, we shall leave D unspecified except by the requirements listed under I. It is comforting to note that the collision states and collision matrix of the theory are uniquely determined by the fields as defined on D_0

† Defined in Section 2–6.

(Refs. 5 and 15). Furthermore, it can be proved that the infinitesimal operators of the Poincaré group, P^{μ} and $M^{\mu\nu}$, have unique self-adjoint extensions when defined only on D_0. We shall see at the beginning of Section 3–2 that there is a domain D_1 larger than D_0 to which the fields can always be extended. For all the practical purposes occurring in Chapter 4 one can identify D with D_1 or D_0.

II. Transformation Law of the Field

The equation

$$U(a, A)\varphi_j(f)U(a, A)^{-1} = \sum S_{jk}(A^{-1})\varphi_k(\{a, A\}f) \quad (3\text{–}4)$$

is valid when each side is applied to any vector in D. Here

$$\{a, A\}f(x) = f(A^{-1}(x - a)). \quad (3\text{–}5)$$

So far, the φ_j are components of some representation of $SL(2,C)$ or the group $SL(2,C)$ combined with reflections, depending on the relativity group. Since, as has already been remarked in Chapter 1, all finite-dimensional representations are sums of irreducible representations, it is natural to regard the components of an irreducible representation as forming a field. Other irreducible representations may then be regarded as different fields occurring in the same theory. However, it is sometimes useful to group together components which transform according to a reducible representation, and regard them as being components of the same field. This is purely a matter of convenience. Our notation in the following will be flexible. When we need to distinguish different sets of components we shall usually use different Greek letters for them.

Examples:

(a) Spin Zero Field

This has one component and if we rewrite (3–4) in the usual unsmeared form, we get

$$U(a, A)\varphi(x)U(a, A)^{-1} = \varphi(\Lambda x + a).$$

(b) Vector Field

This has four components which we write j_μ, $\mu = 0, 1, 2, 3$:

$$U(a, A)j_\mu(x)U(a, A)^{-1} = \Lambda_\mu{}^\nu(A^{-1})j_\nu(\Lambda x + a).$$

(c) Dirac Field ψ

Here again there are four components $\psi_\alpha(x)$, $\alpha = 1, 2, 3, 4$, but the $S(A)$ is the set of 4×4 matrices defined by means of (1–44).

$$U(a, A)\psi_\alpha(x)U(a, A)^{-1} = \sum_{\beta=1}^{4} S(A^{-1})_{\alpha\beta}\psi_\beta(\Lambda x + a).$$

The next assumption is based on the idea that field measurements should commute if they are carried out at points which are space-like separated.

III. Local Commutativity, Sometimes Called Microscopic Causality

If the support of f and the support of g are space-like separated [i.e., if $f(x)g(y) = 0$ for all pairs of points such that $(x - y)^2 \geqslant 0$], then one or the other of

$$[\varphi_j(f), \varphi_k(g)]_\pm \equiv \varphi_j(f)\varphi_k(g) \pm \varphi_k(g)\varphi_j(f) = 0 \quad (3\text{–}6)$$

holds for all j,k when the left-hand side is applied to any vector in D. Similarly,

$$[\varphi_j(f), \varphi_k(g)^*]_\pm = 0.$$

Put in terms of unsmeared fields, this assumption is simply

$$[\varphi_j(x), \varphi_k(y)]_\pm = 0$$

if $(x - y)^2 < 0$, and

$$[\varphi_j(x), \varphi_k^*(y)]_\pm = 0,$$

where $\varphi^*(y)$ is the field that to the test function $g(y)$ gives the operator $[\varphi(\bar{g})]^*$, i.e., $\varphi^*(g) = \varphi(\bar{g})^*$.

For observable fields, we would expect the minus sign to hold. When the + sign holds we shall say that the operators φ_j, φ_k *anti-commute*; in that case bilinear combinations of the fields will commute. The famous theorem on the connection of spin with statistics, proved in Chapter 4, tells us which sign must be taken. (One alternative gives a contradiction!) The fact that anti-commutation relations are of interest is not indicated by any classical analogue; this was one of the discoveries of the founders of our subject.

The preceding assumptions define what we mean by a *field* in a relativistic quantum theory; however they do not yet characterize a *field theory*. For example, in *any* relativistic quantum theory the scalar field $\varphi(x) = c1$, c a constant, satisfies our assumptions, so any relativistic quantum theory has at least a one-parameter family of fields, albeit trivial ones. To be a field theory, a relativistic quantum theory must have enough fields so its states can be uniquely characterized using fields and functions of fields.

In old-fashioned field theory, this requirement was often met by assuming that the fields provided an irreducible set of operators satisfying the canonical commutation relations at a given time:

$$[\varphi_j(\mathbf{x}, t), \pi_k(\mathbf{y}, t)] = i\delta(\mathbf{x} - \mathbf{y})\delta_{jk} \quad (3\text{–}7)$$

(Here j runs over a suitable subset of the indices and the π_j are some combinations of the fields and their space-time derivatives.) However, (3–7) requires that the fields make sense as operators when smeared in $\mathbf{x}$ only, and this is an additional strong assumption which goes beyond our axioms. Furthermore there are hints from examples that, in general, $[\varphi(\mathbf{x}, t), (\partial\varphi/\partial t')(\mathbf{y}, t')]$ has singularities at $t - t' = 0$, even after being smeared in $\mathbf{x}$ and $\mathbf{y}$. In this case it is difficult to give (3–7) a meaning. Thus, one is reluctant to accept canonical commutation relations as an indispensable requirement on a field theory.

Of all the assumptions usually made in canonical quantum field theory, one of the least objectionable is the following: *the smeared fields form an irreducible set of operators in Hilbert space.* Precisely what this means is this:† if B is any bounded operator which satisfies

$$(\Phi, B\varphi_j(f)\,\Psi) = (\varphi_j(f)^*\Phi, B\Psi) \tag{3–8}$$

for all $\Phi, \Psi \in D$, all j and all $f \in \mathscr{S}$, then B is a constant multiple of the identity operator. Notice that (3–8) is a formulation of the condition that B commutes with $\varphi_j(f)$ which avoids assuming that $\varphi_j(f)$ can be defined on vectors $B\Psi$, where $\Psi \in D$. The irreducibility of fields means, roughly, that every operator is a function of the field operators.

There is another notion of commutativity which is commonly used. (See Riesz-Nagy, Ref. 25 of our Chapter II, pp. 301–303.) It is the following: a bounded operator B commutes with a (not necessarily bounded) operator T if (1) B carries the domain, $D(T)$, of T into itself: $BD(T) \subset D(T)$, and (2) for vectors $\Phi \in D(T)$, $BT\Phi = TB\Phi$. A bounded operator which commutes with $\varphi_j(f)$ in this sense surely commutes with $\varphi_j(f)$ in the sense of (3–8).

Interestingly enough, there is an apparently weaker assumption than irreducibility which by virtue of O, I, II, and III ensures irreducibility, and we shall adopt it as our definition.

Definition:

A relativistic quantum theory satisfying axiom O with a field φ_j, $j = 1, \ldots n$ satisfying I, II, and III, is a *field theory* if the vacuum state is *cyclic* for the smeared fields, that is, if polynomials in the smeared field components $P(\varphi_1(f), \varphi_2(g), \ldots)$, when applied to the vacuum state, yield a set D_0 of vectors dense in the Hilbert space of states.

The proof that the cyclicity of the vacuum implies the irreducibility of the field is given in Section 4–2.

† This idea was introduced in Ref. 5.

We have said that the cyclicity of the vacuum is an apparently weaker assumption than irreducibility. For a set of bounded operators, the cautious word "apparently" can be dropped; any non-zero vector is a cyclic vector of an irreducible set of bounded operators invariant under the adjoint operation. (*Proof:* Suppose $\Psi \neq 0$ is *not* cyclic. The subspace $\mathscr{H}_1$, spanned by the $\mathscr{P}\Psi$, where the $\mathscr{P}$ are all polynomials in the bounded operators, is evidently invariant under any of the operators and, by assumption, is not the whole Hilbert space. The projection on $\mathscr{H}_1$ is a non-trivial operator which commutes with all the given bounded operators and contradicts the irreducibility. Therefore Ψ is cyclic.) In our case, in which the relation between D and D_0 is not explicitly known, it is not obvious that the irreducibility of the fields as defined on D implies the cyclicity of Ψ_0. If $D = D_0$ it is not difficult to show that it follows.

Up to this point our assumptions about field theory have made no contact with collision theory. In accordance with the discussion in Section 1–4, we should like to assume

IV. Asymptotic Completeness

$$\mathscr{H} = \mathscr{H}^{\text{in}} = \mathscr{H}^{\text{out}},$$

but in order to do this, we must have some prescription for computing the collision states of all elementary systems of a field theory given the fields. There are several approaches to this problem, but the one which makes closest contact with O, I, II, and III is due to Haag and Ruelle. Ruelle has shown that O, I, II, and III imply the existence of collision states, that is, incoming and outgoing states of one, two, or more particles, provided that the one-particle states can be created by a polynomial in the smeared fields. Then one can formulate axiom IV in terms of these collision states.

It should be mentioned that there need be no connection between the number of fields necessary to describe a field theory and the number of elementary systems it describes. One can have one field and many elementary systems or many fields and only one elementary system. In this book we shall not explain the Haag–Ruelle theory (for a systematic account the reader is referred to the original papers and to Jost's book). We will use only axioms O, I, II, and III, and the assumption that we are dealing with a field theory.

3–2. INDEPENDENCE AND COMPATIBILITY OF THE AXIOMS

There is one class of field theories known to satisfy axioms O, I, II, III, and IV: the theories of free fields of various masses and spins, described heuristically at the end of Section 1–4. These theories will now be described more precisely. That they exist shows that the axioms are compatible, but in a physically trivial way since the particles they describe do not interact.

In all the free field theories the total number of particles is an integral of motion and the Hilbert space of states is written as a direct sum,

$$\mathscr{H} = \bigoplus_{n=0}^{\infty} \mathscr{H}^{(n)}, \tag{3–9}$$

where $\mathscr{H}^{(n)}$ is the subspace of those states with exactly n particles. This means that a vector, Φ, of $\mathscr{H}$ is given by a sequence

$$\{\Phi^{(0)}, \Phi^{(1)}, \Phi^{(2)}, \ldots\} \tag{3–10}$$

of vectors with $\Phi^{(j)} \in \mathscr{H}^{(j)}$, and the scalar product in $\mathscr{H}$ is

$$(\Phi, \Psi) = \sum_{n=0}^{\infty} (\Phi^{(n)}, \Psi^{(n)}), \tag{3–11}$$

where the scalar product $(\Phi^{(n)}, \Psi^{(n)})$ is in $\mathscr{H}^{(n)}$. Only those sequences are in $\mathscr{H}$ which satisfy

$$(\Phi, \Phi) = \sum_{n=0}^{\infty} \|\Phi^{(n)}\|^2 < \infty. \tag{3–12}$$

For all spins, $\mathscr{H}^{(0)}$ is the one-dimensional Hilbert space of complex numbers. For spin s, the Hilbert space $\mathscr{H}^{(n)}$ is the space of all square-integrable functions of the arguments $p_1\alpha_1, \ldots, p_n\alpha_n$, symmetric if s is integral and anti-symmetric if s is half-odd-integral. Here α_j stands for a group of undotted spinor indices like those in (1–58). The scalar product in $\mathscr{H}^{(n)}$ is

$$(\Phi^{(n)}, \Psi^{(n)}) = \int \cdots \int d\Omega_m(p_1) \ldots d\Omega_m(p_n) \sum_{\substack{\alpha_1 \ldots \alpha_n \\ \beta_1 \ldots \beta_n}} \overline{\Phi^{(n)}(p_1\alpha_1, \ldots p_n\alpha_n)}$$

$$\times \prod_{j=1}^{n} \mathscr{D}^{(s,0)}(\tilde{p}_j/m)_{\alpha_j\beta_j} \Psi^{(n)}(p_1\beta_1, \ldots p_n\beta_n) \tag{3–13}$$

and the field operator φ_α is defined by

$$\varphi_\alpha(f)\Psi^{(n)}(p_1\alpha_1, \ldots p_n\alpha_n)$$

$$= \sqrt{\pi}\Big\{\sqrt{n+1} \int d\Omega_m(p) \tilde{f}(p) \Psi^{(n+1)}(p\alpha, p_1\alpha_1, \ldots p_n\alpha_n)$$

$$+ \frac{1}{\sqrt{n}} \sum_{j=1}^{n} (-1)^{2s(j+1)} \tilde{f}(-p_j) \mathscr{D}^{(s,0)}(\zeta)_{\alpha\alpha_j}$$

$$\times \Psi^{(n-1)}(p_1\alpha_1, \ldots \hat{p}_j\hat{\alpha}_j, \ldots p_n\alpha_n)\Big\}. \tag{3–14}$$

Here ˆ over a letter means omit it, and $\tilde{f}$ is defined by

$$\tilde{f}(p) = \frac{1}{(2\pi)^2} \int e^{-ip\cdot x} f(x)\, dx. \tag{3–15}$$

The representation of the inhomogeneous $SL(2,C)$ is here

$$\begin{aligned}(U(a, A)\Psi)^{(n)}(p_1\alpha_1, \ldots p_n\alpha_n) &= \exp\left[i\left(\sum_{j=1}^{n} p_j\right)\cdot a\right] \sum_{(\beta)} \\ &\times \prod_{j=1}^{n} \mathscr{D}^{(s,0)}(A)_{\alpha_j\beta_j} \Psi^{(n)}(A^{-1}p_1\beta_1, \ldots A^{-1}p_n\beta_n). \end{aligned} \tag{3–16}$$

For the domain D, one can take, for example, those Ψ for which $\Psi^{(n)}$ vanishes for all sufficiently large n and is the restriction to the hyperboloids $p_j^2 = m^2$ of an infinitely differentiable function in $\mathscr{S}(\mathbf{R}^{4n})$. It is a straightforward task to verify 0, I, II, III, and IV for the theory given by these definitions, an exercise we strongly recommend to the reader. For the case of spin zero, the representation of the inhomogeneous $SL(2,C)$ is just that illustrated in Figure 1–3.

There are fields different from free fields, which can be constructed in terms of free fields. For example, consider the fields which are indicated in terms of the scalar field, φ, by the limiting procedure

$$:D^\alpha\varphi(x)D^\beta\varphi(x): = \lim_{x_1,x_2\to x} [D^\alpha\varphi(x_1)D^\beta\varphi(x_2) - (\Psi_0, D^\alpha\varphi(x_1)D^\beta\varphi(x_2)\Psi_0)]$$

and

$$\begin{aligned} :D^\alpha\varphi(x)D^\beta\varphi(x)D^\gamma\varphi(x): &= \lim_{x_1,x_2,x_3\to x} [D^\alpha\varphi(x_1)D^\beta\varphi(x_2)D^\gamma\varphi(x_3) \\ &\quad - (\Psi_0, D^\alpha\varphi(x_1)D^\beta\varphi(x_2)\Psi_0)D^\gamma\varphi(x_3) \\ &\quad - (\Psi_0, D^\alpha\varphi(x_1)D^\gamma\varphi(x_3)\Psi_0)D^\beta\varphi(x_2) \\ &\quad - (\Psi_0, D^\beta\varphi(x_2)D^\gamma\varphi(x_3)\Psi_0)D^\alpha\varphi(x_1)] \end{aligned} \tag{3–17}$$

and analogously for higher degrees in φ. [Recall the definition of D^α in (2–5).] It is not difficult to show, by using formula (3–14), that the right-hand sides of these expressions indeed define fields satisfying 0, I, II, and III with the same domain D as that of the free field, and the same representation $U(a, A)$. They have transformation laws appropriate to the tensor indices α, β, or α, β, γ, etc., respectively. By contracting the indices we get scalar fields, for example,

$$\psi(x) = \varphi(x) + :\frac{\partial}{\partial x_\mu}\varphi(x)\frac{\partial}{\partial x^\mu}\varphi(x):. \tag{3–18}$$

Such fields are sometimes called Wick polynomials. The particular example (3–18) satisfies axiom IV in addition to O, I, II, and III, but it gives the same S-matrix as the free field. That is what always happens for a field theory defined by a Wick polynomial, as will be seen in Chapter 4.

Another class of examples of fields can be obtained by replacing the $d\Omega_m(p)$ in (3–13) and (3–14) by $\int_0^\infty \rho(a)\, da\, d\Omega_a(p)$, where $\rho(a)$ is a non-negative weight. Unless $\rho(a) = \delta(a - m)$, this gives a representation of the Lorentz group different from that associated with the free field of mass m. Such fields are called *generalized free fields*. It is proved in Ref. 4 that axiom IV is not satisfied for certain choices of ρ. This shows that IV is independent of O, I, II, and III.

These examples show that there are field theories whose fields satisfy O, I, II, and III for certain representations $\{a, A\} \rightarrow U(a, A)$. Examples of representations also exist which have no non-trivial fields having the vacuum as cyclic vector. For example, that is the case if the representation contains no masses above some limit. This is a consequence of a theorem proved in Ref. 15, which says that the energy-momentum spectrum of a field theory must be additive; that is, if p_1 and p_2 are in the spectrum so is $p_1 + p_2$.

It is fairly obvious that III is independent of O, I, and II since it is easy to write down non-local fields satisfying O, I, and II. In fact it has been repeatedly suggested that local commutativity (axiom III) is too strong an assumption, since there is no evidence for (or against) the simultaneous measurability of fields at points of space-time which are separated by very small distances, say 10^{-16} cm. In this connection Theorem 4–1 is relevant. If III is replaced by commutativity at large space-like distances, say $(x - y)^2 < -\ell^2$ (thus introducing a fundamental length), then III follows. Thus if III is to be essentially weakened the commutator must be allowed to be different from zero at all space-like distances.

It has been suggested that the irreducibility assumption in the definition of a field theory should be strengthened by the *time slice axiom* as follows: if T is a set in space-time lying between two space-like surfaces, then operators of the form $\varphi(f)$ should form an irreducible set, where $f = 0$ outside T. This is a mathematically precise substitute for the ill-defined requirement of irreducibility of the fields and their derivatives at a given time. We shall not analyze the time slice axiom here. We note that it is stronger than assuming the cyclicity of the vacuum, as has been shown in Ref. 3.

The examples given in this section are trivial in the sense that, though they satisfy O, I, II, and III, they yield a trivial collision theory.

As was already mentioned in the introduction, it is a central problem of the theory at the moment to find examples which are non-trivial in this sense.

3–3. PROPERTIES OF THE VACUUM EXPECTATION VALUES

The object of this section is to discuss the quantities

$$(\Psi_0, \varphi_1(x_1)\varphi_2(x_2)\ldots\varphi_n(x_n)\Psi_0) \tag{3–19}$$

where φ_j, $j = 1,\ldots n$ is a component of an irreducible tensor. We shall refer to such expressions as *vacuum expectation values.* Two of the φ_j might be components of different fields or could be the same component of the same field or could be hermitian conjugate to each other. In other words, we consider all such quantities, for all combinations of the arbitrary labels, and all permutations of the indices.

To give a meaning to such a symbol, we note that axiom I implies that

$$(\Psi_0, \varphi_1(f_1)\varphi_2(f_2)\ldots\varphi_n(f_n)\Psi_0) \tag{3–20}$$

exists and is a separately continuous multilinear functional of the arguments $f_1,\ldots f_n$ as they vary over $\mathscr{S}(\mathbf{R}^4)$. It follows from the nuclear theorem (Theorem 2–1) that this functional can be uniquely extended to be a tempered distribution of the n four-vectors $x_1,\ldots x_n$. It is this distribution that (3–19) symbolizes, and which, for the particular choice of field-components $\varphi_1,\ldots\varphi_n$ (from among all components) under discussion at a given moment, we shall denote by $\mathscr{W}$:

$$\mathscr{W}(x_1, x_2,\ldots x_n) = (\Psi_0, \varphi_1(x_1)\varphi_2(x_2)\ldots\varphi_n(x_n)\Psi_0). \tag{3–21}$$

With the same choice of labels we shall denote by $\mathscr{W}_\pi$ the "permuted" vacuum expectation value

$$\mathscr{W}_\pi(x_1, x_2,\ldots x_n) = (\Psi_0, \varphi_{i_1}(x_{i_1})\varphi_{i_2}(x_{i_2})\ldots\varphi_{i_n}(x_{i_n})\Psi_0), \tag{3–22}$$

where π is the permutation $(1, 2,\ldots n) \to (i_1, i_2,\ldots i_n)$. Clearly if all the components φ_j are different there are $n!$ different $\mathscr{W}_\pi$ for each choice of components, and fewer if some coincide.

The argument used above to give a meaning to (3–19) as a distribution can also be used to make sense of expressions of the form

$$\Psi = \int dx_1\ldots dx_k f(x_1,\ldots x_k)\varphi_1(x_1)\ldots\varphi_k(x_k)\Psi_0. \tag{3–23}$$

We can find a sequence of functions $\{f_J\}$, with

$$f_J(x_1,\ldots x_k) = \sum_{j=1}^{J} f_1{}^j(x_1)\ldots f_k{}^j(x_k)$$

and $f_i^j \in \mathscr{S}(\mathbf{R}^4)$ such that $f_J \to f$ in $\mathscr{S}(\mathbf{R}^{4k})$ as $J \to \infty$. Then the sequence of states

$$\Psi_J = \sum_{j=1}^{J} \varphi_1(f_1{}^j)\varphi_2(f_2{}^j)\ldots\varphi_k(f_k{}^j)\Psi_0 \tag{3–24}$$

converges in norm, and we take its limit as the definition of Ψ in (3–23). To prove the convergence of (3–24), consider

$$\begin{aligned}\|\Psi_m - \Psi_n\|^2 &= \left(\sum_{j=m}^{j=n}\varphi_1(f_1{}^j)\ldots\varphi_k(f_k{}^j)\Psi_0, \sum_{j=m}^{j=n}\varphi_1(f_1{}^j)\ldots\varphi_k(f_k{}^j)\Psi_0\right)\\ &= \int \mathscr{W}(y_k,\ldots y_1, x_1,\ldots x_k)\\ &\qquad\qquad \times (\bar{f}_n - \bar{f}_m)(f_n - f_m)\,dx_1\ldots dx_k\,dy_1\ldots dy_k,\end{aligned}$$

where $\mathscr{W}$ corresponds to the fields $\varphi_k^*\ldots\varphi_1^*\varphi_1\ldots\varphi_k$ in that order. Now the function $(\bar{f}_n - \bar{f}_m)(f_n - f_m)$ tends to zero in $\mathscr{S}$, and since $\mathscr{W}$ is a tempered distribution, $\|\Psi_m - \Psi_n\| \to 0$. Since $\mathscr{H}$ is a complete space,† $\{\Psi_m\}$ has a limit state Ψ.

The set of all vectors symbolized by (3–23) will be denoted D_1. Clearly, any $\varphi_j(f)$ can be defined by continuity on D_1,

$$\varphi_j(f)\Psi = \int dx_1\ldots dx_k\,dx\,f(x)f(x_1,\ldots x_k)\varphi_j(x)\varphi_1(x_1)\ldots\varphi_k(x_k)\Psi_0.$$

It is easy to verify that the knowledge of the φ_j on the domain D_0 and the validity of I, II, and III there imply I, II, and III on D_1.

We express our assertions about the vacuum expectation values in several theorems. The first is

Theorem 3–1

(*a*) (*Relativistic Transformation Law*). Suppose $\varphi_\alpha, \psi_\beta, \ldots \chi_\gamma$ are n fields which transform according to irreducible representations of $SL(2,C)$, namely

$$U(a, A)\varphi_\alpha(x)U(a, A)^{-1} = \sum_{\alpha'} S^{(\varphi)}_{\alpha\alpha'}(A^{-1})\varphi_{\alpha'}(Ax + a) \tag{3–25}$$

[$\psi,\ldots\chi$ transforming with $S^{(\psi)},\ldots S^{(\chi)}$ respectively.] Then the vacuum expectation values for $n = 1, 2, 3,\ldots$ are tempered distributions which transform according to

$$\begin{aligned}&\sum_{\alpha',\beta',\ldots\gamma'} S_{\alpha\alpha'}{}^{(\varphi)}(A)S_{\beta\beta'}{}^{(\psi)}(A)\ldots S_{\gamma\gamma'}{}^{(\chi)}(A)\\ &\qquad\qquad \times (\Psi_0, \varphi_{\alpha'}(x_1)\psi_{\beta'}(x_2)\ldots\chi_{\gamma'}(x_n)\Psi_0)\\ &= (\Psi_0, \varphi_\alpha(Ax_1 + a)\psi_\beta(Ax_2 + a)\ldots\chi_\gamma(Ax_n + a)\Psi_0).\end{aligned} \tag{3–26}$$

† See Section 2–6.

Proof:

As remarked earlier, the vacuum expectation values are tempered distributions, by the nuclear theorem. Using (3–25) with A replaced by A^{-1}, and the property $U(a, A)\Psi_0 = \Psi_0$ [Eq. (3–1)], (3–26) follows immediately. ▌

Theorem 3–2

Suppose $\varphi_1, \varphi_2, \ldots \varphi_n$ are any components of any fields, and $\mathscr{W}$ is defined as in (3–21). Then

(*b*) (*Spectral Conditions*). There are tempered distributions $W(\xi_1, \ldots \xi_{n-1})$ depending on the relative coordinates

$$\xi_j = x_j - x_{j+1} \qquad j = 1, 2, \ldots n - 1 \tag{3–27}$$

which satisfy

$$\mathscr{W}(x_1, \ldots x_n) = W(\xi_1, \xi_2, \ldots \xi_{n-1}). \tag{3–28}$$

The Fourier transforms of $\mathscr{W}$ and W are tempered distributions defined by

$$\tilde{\mathscr{W}}(p_1, p_2, \ldots p_n) = \int \exp\left(i \sum_{j=1}^{n} p_j x_j\right) \mathscr{W}(x_1, \ldots x_n)\, dx_1 \ldots dx_n \tag{3–29}$$

$$\tilde{W}(q_1, \ldots q_{n-1}) = \int \exp\left(i \sum_{j=1}^{n-1} q_j \xi_j\right) W(\xi_1, \xi_2, \ldots \xi_{n-1})\, d\xi_1 \ldots d\xi_{n-1} \tag{3–30}$$

and are related by

$$\tilde{\mathscr{W}}(p_1, \ldots p_n) = (2\pi)^4\, \delta\left(\sum_{j=1}^{n} p_j\right) \tilde{W}(p_1, p_1 + p_2, \ldots p_1 + p_2 + \cdots + p_{n-1}) \tag{3–31}$$

Further we have

$$\tilde{W}(q_1, \ldots q_{n-1}) = 0 \tag{3–32}$$

if any q is not in the energy momentum spectrum of the states.

(*c*) (*Hermiticity Conditions*)

$$(\Psi_0, \varphi_1(x_1)\varphi_2(x_2) \ldots \varphi_n(x_n)\Psi_0) = \overline{(\Psi_0, \varphi_n^*(x_n) \ldots \varphi_2^*(x_2)\varphi_1^*(x_1)\Psi_0)}. \tag{3–33}$$

(*d*) (*Local Commutativity Conditions*)

$$\mathscr{W}(x_1, x_2, \ldots x_{j+1}, x_j, \ldots x_n) = (-1)^m \mathscr{W}(x_1, x_2, \ldots x_n), \qquad j = 1, \ldots n-1 \quad (3\text{–}34)$$

if the difference $x_j - x_{j+1}$ is space-like, where m is zero if φ_j and φ_{j+1} commute, and one if they anti-commute.

Proof:

The argument given in Section 2–1 shows that translation invariance implies that $\mathscr{W}$ depends only on the differences $\xi_1, \ldots \xi_{n-1}$, or more precisely, that there exists a tempered distribution W satisfying (3–28). Clearly

$$\int dx_1 \ldots dx_n \exp\left(i \sum_{j=1}^{n} p_j x_j\right) \mathscr{W}(x_1, x_2, \ldots x_n)$$

$$= \int dx_1 \ldots dx_n \exp\{i[p_1(x_1 - x_2) + (p_1 + p_2)(x_2 - x_3) + \cdots + (p_1 + \cdots + p_{n-1})(x_{n-1} - x_n)]\}$$

$$\times \exp\left[\left(i \sum_{j=1}^{n} p_j\right) x_n\right] W(x_1 - x_2, \ldots x_{n-1} - x_n)$$

$$= (2\pi)^4 \delta\left(\sum_{j=1}^{n} p_j\right) \tilde{W}(p_1, p_1 + p_2, \ldots p_1 + p_2 + \cdots + p_{n-1}).$$

As explained in Section 2–6, for any two states Φ, Ψ, we have

$$\int e^{-ipa}\, da(\Phi, U(a, 1)\Psi) = 0,$$

unless p lies in the energy-momentum spectrum of the states. Therefore

$$\int e^{ipa}\, da(\Psi_0, \varphi_1(f_1) \ldots \varphi_j(f_j) U(-a, 1) \varphi_{j+1}(f_{j+1}) \ldots \varphi_n(f_n)\Psi_0) = 0,$$

which implies

$$\int e^{ipa} W(\xi_1, \ldots \xi_{j-1}, \xi_j + a, \xi_{j+1}, \ldots \xi_{n-1})\, da = 0, j = 1, 2, \ldots n$$

for p not in the physical spectrum, that is, $\tilde{W}(q_1, \ldots q_{n-1}) = 0$ unless each q_j lies in the physical spectrum.

(*c*) This relation follows immediately from

$$(\Psi_0, \varphi_1(f_1) \ldots \varphi_n(f_n)\Psi_0) = \overline{[(\Psi_0, (\varphi_n(f_n))^* \ldots (\varphi_1(f_1))^* \Psi_0)]}$$

and the fact that we can continue this relation from test functions of the form $f_1(x_1) f_2(x_2) \ldots f_n(x_n)$ to all of $\mathscr{S}(\mathbf{R}^{4n})$.

(d) This relation follows immediately from axiom III, again with the use of the extension from product test functions to all test functions of $\mathscr{S}(\mathbf{R}^{4n})$. ∎

The next property follows from the positive definiteness of the scalar product in Hilbert space.

Theorem 3–3

(e) (*Positive Definiteness Conditions*). For any sequence $\{f_j\}$ of test functions, $f_j \in \mathscr{S}(\mathbf{R}^{4j})$, with $f_j = 0$ except for a finite number of j, the vacuum expectation values satisfy the inequalities

$$\sum_{j,k=0}^{\infty} \int \cdots \int \bar{f}_j(x_1, \ldots x_j) \mathscr{W}_{jk}(x_j, x_{j-1}, \ldots x_1, y_1, \ldots y_k) \times f_k(y_1, \ldots y_k)\, dx_1 \ldots dx_j\, dy_1 \ldots dy_k \geqslant 0. \quad (3\text{–}35)$$

In (3–35) $\mathscr{W}_{jk}$ represents

$$(\Psi_0, \varphi_{jj}^*(x_j) \ldots \varphi_{j1}^*(x_1) \varphi_{k1}(y_1) \ldots \varphi_{kk}(y_k) \Psi_0),$$

where the field components labeled by jk can be any selection from among all components in the theory. Moreover, if, for any sequence $f_0, f_1, \ldots$ we have the equality sign in (3–35), then† it must give 0 also when the sequence, $\{f_j\}$, is replaced by a sequence $\{g_j\}$ where

$$g_0 = 0, \quad g_1 = g(x_1) f_0, \quad g_2 = g(x_1) f_1(x_2), \quad g_3 = g(x_1) f_2(x_2, x_3) \ldots \quad (3\text{–}36)$$

for any test function g.

Proof:

The inequalities (3–35) express the fact that the norm of the state

$$\Psi = f_0 \Psi_0 + \varphi_{11}(f_1)\Psi_0 + \int \varphi_{21}(x_1)\varphi_{22}(x_2) f_2(x_1, x_2)\, dx_1\, dx_2 \Psi_0 + \cdots + \int \varphi_{j1}(x_1)\varphi_{j2}(x_2) \ldots \varphi_{jj}(x_j) f_j(x_1, \ldots x_j)\, dx_1 \ldots dx_j \Psi_0 + \cdots$$

is non-negative. If the norm is 0, $\Psi = 0$, and so $\varphi_\alpha(g)\Psi = 0$ for any component φ_α and any test function g. Hence the expression (3–35) must give 0 also for the sequence (3–36). ∎

As we shall see in the next section, the conditions (a) through (e) are sufficient to guarantee that a set of tempered distributions are the vacuum expectation values of some field theory satisfying O, I, II, and III, with the exception of the requirement that the vacuum be unique. The next theorem gives an additional condition which ensures this property, as we shall see in Section 3–4.

† It is shown on p. 121 that this follows from (3–35) and the other axioms.

Theorem 3–4 (*Cluster Decomposition Property*)

If a is a space-like vector, then

$$\mathscr{W}(x_1,\ldots x_j, x_{j+1} + \lambda a, x_{j+2} + \lambda a,\ldots x_n + \lambda a)$$
$$\rightarrow \mathscr{W}(x_1,\ldots x_j)\mathscr{W}(x_{j+1},\ldots x_n) \quad (3\text{–}37)$$

as $\lambda \rightarrow \infty$, in the sense of convergence in $\mathscr{S}'$.

Remark:

We shall prove the theorem only for a theory in which there is no state other than the vacuum having a mass less than some positive threshold M (which may be as small as we like). Such theories will be said to have a *mass gap*. If there are mass zero particles in the theory there is no mass gap, but (3–37) still holds; the proof is more technical in this case. Equation (3–37) can also be proved without assuming axiom III. See Refs. 5, 7, 9, 10, and 11 for further information.

Proof:

The proof is based on a very simple argument due to D. Ruelle. It exploits the fact that for large λ the operators occurring in (3–37) can be rewritten in the order $x_{j+1} + \lambda a,\ldots x_n + \lambda a, x_1,\ldots x_j$ with at most a change in sign in the function. This reversal of order has the effect of reversing the sign of the momentum conjugate to the difference variable $x_j - x_{j+1}$. The change of sign makes it possible to use the spectral condition in this momentum vector to get the theorem. The details are as follows.

For convenience, choose $a^2 = -1$. This can be done without loss of generality by changing the scale of λ. The Fourier transforms of the distributions

$$F_1 = \mathscr{W}(x_1,\ldots x_j, x_{j+1} + \lambda a,\ldots x_n + \lambda a) - \mathscr{W}(x_1,\ldots x_j)\mathscr{W}(x_{j+1},\ldots x_n)$$
$$F_2 = \mathscr{W}(x_{j+1} + \lambda a,\ldots x_n + \lambda a, x_1,\ldots x_j) - \mathscr{W}(x_1,\ldots x_j)\mathscr{W}(x_{j+1},\ldots x_n)$$

are zero unless $P = p_1 + \cdots + p_j$ lies in the forward and backward cones, respectively, by property (*b*). If there is a unique vacuum state, Ψ_0, then the product $\mathscr{W}(x_1,\ldots x_j)\mathscr{W}(x_{j+1},\ldots x_n)$ cancels the contribution from this state at the origin $P^\mu = 0$, and we get $\tilde{F}_1 \pm \tilde{F}_2 = 0$ unless $P^2 \geqslant M^2$. For fixed $x_1,\ldots x_n$, F_1 and F_2 coincide up to a sign for large enough λ, by locality. From Section 2–1, $F_1 \pm F_2$ is a finite sum of derivatives of a polynomially bounded continuous function, G say. Thus, for any test function $h \in \mathscr{S}$,

$$\int (F_1 \pm F_2)(x_1,\ldots x_n)h(x_1,\ldots x_n)\, dx_1\ldots dx_n$$
$$= \int D^m G(\lambda, x_1,\ldots x_n)h(x_1,\ldots x_n)\, dx_1\ldots dx_n.$$

Now G is polynomially bounded in all variables, i.e.,

$$|G| \leqslant G_0\lambda^N R^Q \qquad \text{for sufficiently large } \lambda \text{ and } R,$$

where by R we mean the *euclidean* norm in $\mathbf{R}^{4n}$ over which the integral runs, i.e.,

$$R^2 = \sum_j [(x_j{}^0)^2 + (\mathbf{x}_j)^2].$$

Using local commutativity we can and will choose the $\pm$ sign so that $D^m G = 0$ for large λ. We shall need a more quantitative form of this statement: $D^m G = 0$ for $R < R_0$, where R_0 is a positive multiple of λ. To determine R_0 note that

$$\begin{aligned}(x_i - x_k - \lambda a)^2 \\ &= (x_i{}^0 - x_k{}^0)^2 - (\mathbf{x}_i - \mathbf{x}_k)^2 - \lambda^2 - 2\lambda\left(a^0(x_i{}^0 - x_k{}^0) - \mathbf{a}\cdot(\mathbf{x}_i - \mathbf{x}_k)\right) \\ &< |x_i{}^0|^2 + |x_k{}^0|^2 + 2|x_i{}^0|\,|x_k{}^0| - \lambda^2 \\ &\qquad + 2\lambda\left(|a^0|\,(|x_i{}^0| + |x_k{}^0|) + |\mathbf{a}|\,(|\mathbf{x}_i| + |\mathbf{x}_k|)\right)\end{aligned}$$

so that when

$$R^2 = \sum_{i=1}^{n} [(x_i{}^0)^2 + \mathbf{x}_i{}^2] < R_0{}^2$$

we have

$$(x_i - x_k - \lambda a)^2 < 4R_0{}^2 - \lambda^2 + 4\lambda R_0(|a^0| + |\mathbf{a}|).$$

The right-hand side is <0 for all positive λ if R_0 is chosen as a sufficiently small multiple of λ, say $R_0 = \frac{1}{8}[|a^0| + |\mathbf{a}|]^{-1}\lambda$. From this estimate follows, for any $\varepsilon > 0$ and sufficiently small,

$$\begin{aligned}&\left|\int D^m G(\lambda, x_1, \ldots x_n) h(x_1, \ldots x_n)\, dx_1 \ldots dx_n\right| \\ &= \left|\int_{R > R_0 - \varepsilon} D^m G(\lambda, x_1, \ldots x_n) h(x_1, \ldots x_n)\, dx_1 \ldots dx_n\right| \\ &= \left|\int_{R > R_0 - \varepsilon} G(\lambda, x_1, \ldots x_n) D^m h(x_1, \ldots x_n)\, dx_1 \ldots dx_n\right| \\ &\leqslant \int_{R > R_0 - \varepsilon} |G|\,|D^m h|\, R^{4n-1}\, dR\, d\omega\end{aligned}$$

where in the last expression we have passed to polar coordinates. Since for any q, the inequality $|D^m h| < cR^{-q}$ holds for some c and all sufficiently large R, the right-hand side of the preceding inequality is bounded by

$$\int_{R > R_0} G_0 c \lambda^N R^{Q-q+4n-1}\, dR\, d\omega$$

for all sufficiently large λ. For sufficiently large q this integral converges and shows that

$$|\int (D^m G)h| < \frac{c_1}{\lambda^p}$$

for all p, some c_1, and all sufficiently large λ. This is what is meant by

$$\lambda^p(F_1 \pm F_2)(x_1,\ldots x_j, x_{j+1} + \lambda a,\ldots x_n + \lambda a) \to 0 \qquad \text{as } \lambda \to \infty$$

in the sense of convergence in $\mathscr{S}'$.

The final step of the proof is to show that this equation holds for F_1 and F_2 separately, and here is where the spectral condition is used. Replace the h of the preceding argument by a new h_1 which is defined by $\tilde{h}_1 = \theta\tilde{h}$ where θ is an infinitely differentiable function of the variable $P = \sum_{k=1}^{j} p_k$, equal to 1 for $P^2 \geqslant M^2$, $P^0 > 0$, and 0 for $P^0 \leqslant 0$. Clearly $h_1 \in \mathscr{S}$ so the preceding argument works equally well for it. But $\int F_2 h_1\, dx_1\ldots dx_n = 0$ and $\int F_1 h_1\, dx_1\ldots dx_n = \int F_1 h\, dx_1\ldots dx_n$, so the required result (3–37) follows. ▌

The physical meaning of the cluster decomposition property is that, when two systems at points x and y become separated by a large space-like distance, the interaction between them falls off to zero. The method of proof given here shows that if there is a mass gap, the limit converges faster than any inverse power $1/\lambda$. It can be shown to decrease exponentially, the damping factor depending on the threshold mass M. If there are zero-mass particles in the theory the limit may go as slowly as $1/\lambda^2$; this is just the Coulomb force!

If the fields have doubled components so that (1–52) makes sense, we can formulate the requirement that $U(I_s)$ and $U(C)$ are unitary operators by substituting these transformation laws on each component of the product of fields entering in the state (3–23), and requiring that all scalar products be left invariant. This leads to further properties of the $\mathscr{W}$'s. For example, if charge conjugation defines a symmetry, then

$$(\Psi_0, \varphi_{(\alpha)(\dot\beta)}(x_1)\ldots\psi_{(\mu)(\dot\nu)}(x_n)\Psi_0) = (\Psi_0, \varphi^*_{(\dot\alpha)(\beta)}(x_1)\ldots\psi^*_{(\dot\mu)(\nu)}(x_n)\Psi_0) \qquad (3\text{–}38)$$

holds for all products in all orders. Equation (3–38) is a further restriction on the $\mathscr{W}$'s, and should not be confused with the rule for taking the hermitian conjugate, (3–33). If Θ is a symmetry of the theory we get the relations

$$(\Psi_0, \varphi_{(\alpha)(\dot\beta)}(x_1)\ldots\psi_{(\mu)(\dot\nu)}(x_n)\Psi_0)$$
$$= i^F(-1)^J\overline{(\Psi_0, \varphi^*_{(\dot\alpha)(\beta)}(-x_1)\ldots\psi^*_{(\dot\mu)(\nu)}(-x_n)\Psi_0)} \qquad (3\text{–}39)$$

from (1–53), since Θ is anti-unitary, where F is the total number of half-odd integer spin fields and J is the total number of undotted indices in the collections $(\alpha),\ldots(\mu)$.

In the same way, equations similar to (3–38) and (3–39) can be found if I_s and I_t define symmetries.

We now introduce an important technique in the study of local fields; we relate the vacuum expectation values to holomorphic functions.

Theorem 3–5

The $\mathscr{W}$ and W are boundary values of holomorphic functions in the following sense. There exist holomorphic functions $\mathscr{W}$ and W, such that

$$\mathscr{W}(z_1,\ldots z_n) = W(z_1 - z_2, z_2 - z_3,\ldots z_{n-1} - z_n),$$

and W is holomorphic in the variables $\zeta_j = \xi_j - i\eta_j = z_j - z_{j+1}$, $j = 1,\ldots n$, the domain of holomorphy being the tube $\{\zeta_1,\ldots\zeta_{n-1} \mid \eta_j \in V_+\}$, and

$$W(\xi_1,\ldots\xi_{n-1}) = \lim_{\eta_1\ldots\eta_{n-1}\to 0} W(\xi_1 - i\eta_1,\ldots\xi_{n-1} - i\eta_{n-1}),$$

where the limit is to be understood in the sense of convergence in $\mathscr{S}'$. Furthermore, W is polynomially bounded† in $\mathscr{T}_{n-1}$. The holomorphic functions W possess a single-valued continuation into the extended tube $\mathscr{T}'_{n-1}$.

Proof:

Since $\tilde{W}(q_1,\ldots q_{n-1}) = 0$ unless each q lies in the forward cone, the Laplace transform

$$W(\xi_1 - i\eta_1,\ldots\xi_{n-1} - i\eta_{n-1})$$
$$= (2\pi)^{-4(n-1)}\int\ldots\int dq_1\ldots dq_{n-1} \exp[-i\Sigma(\xi_j - i\eta_j)\cdot q_j]\tilde{W}(q_1,\ldots q_{n-1})$$

is a holomorphic function, by Theorem 2–8. The other assertions, except for the last, follow from the same theorem.

We now make use of Lorentz invariance to show that each vacuum expectation value has a one-valued continuation into a much larger region, the extended tube $\mathscr{T}'_{n-1}$. Consider (3–26):

$$\sum_{\alpha'\beta'\gamma'} S^{(\varphi)}_{\alpha\alpha'}(A)S^{(\psi)}_{\beta\beta'}(A)\ldots S^{(\chi)}_{\gamma\gamma'}(A)(\Psi_0, \varphi_{\alpha'}(x_1)\psi_{\beta'}(x_2)\ldots\chi_{\gamma'}(x_n)\Psi_0)$$
$$= (\Psi_0, \varphi_\alpha(Ax_1 + a)\psi_\beta(Ax_2 + a)\ldots\chi_\gamma(Ax_n + a)\Psi_0).$$

† Defined in Section 2–3.

The left-hand side has a continuation for complex Λ, providing a continuation of the right-hand side by

$$\sum_{\alpha'\beta'\gamma'} (\Psi_0, \varphi_{\alpha'}(x_1)\psi_{\beta'}(x_2)\ldots\chi_{\gamma'}(x_n)\Psi_0)$$
$$\times S^{(\varphi)}_{\alpha\alpha'}(A, B)S^{(\psi)}_{\beta\beta'}(A, B)\ldots S^{(\chi)}_{\gamma\gamma'}(A, B)$$
$$= (\Psi_0, \varphi_\alpha(\Lambda(A, B)x_1 + a)\psi_\beta(\Lambda(A, B)x_2 + a)\ldots$$
$$\chi_\gamma(\Lambda(A, B)x_n + a)\Psi_0).$$

This continuation defines the right-hand side for all points $\Lambda(A, B)x_j$ with $x_j - x_{j+1} \in \mathscr{T}_{n-1}$, as the A, B vary over $SL(2,C) \otimes SL(2,C)$. The continuation is single-valued, by Theorem 2–11. Thus $W(\zeta_1, \ldots \zeta_{n-1})$ is continuable into $\mathscr{T}'_{n-1}$. ▮

An important special case is $\Lambda = -1$, which is connected in the complex Lorentz group to $\Lambda = 1$. By (1–27) $S^{(\varphi)}(-1, 1) = (-1)^j$, where j is the number of undotted indices in the field φ. Hence, if J is the total number of undotted indices in the fields appearing in the W, we get

$$W(\xi_1, \ldots \xi_{n-1}) = (-1)^J W(-\xi_1, -\xi_2, \ldots - \xi_{n-1}), \qquad (3\text{–}40)$$

which holds at all points of holomorphy. It should be emphasized that (3–40) holds, even if P, C, and T do not define symmetries, since it follows from invariance under $\mathscr{P}^\uparrow_+$ alone, combined with the hypothesis about the mass spectrum of states.

The final theorem relates the function W to the holomorphic function W_π, defined analogously for a permutation of the fields.

Theorem 3–6

Suppose that $W(\xi_1, \xi_2, \ldots \xi_{n-1}) = (\Psi_0, \varphi_1(x_1)\ldots\varphi_n(x_n)\Psi_0)$, $\xi_j = x_j - x_{j+1}$, and

$$W_\pi(\xi_1, \xi_2, \ldots \xi_{n-1}) = (\Psi_0, \varphi_{i_1}(x_{i_1})\varphi_{i_2}(x_{i_2})\ldots\varphi_{i_n}(x_{i_n})\Psi_0),$$

where π is the permutation $(1, 2, \ldots n) \to (i_1, i_2, \ldots i_n)$; and suppose W and W_π are the holomorphic functions given by the previous theorem. Then W and W_π continue one another as one holomorphic function.

Proof:

Suppose $x_1, x_2, \ldots x_n$ are such that all the differences $x_i - x_j$ are space-like.† Then W and W_π coincide in a real neighborhood of this point, by local commutativity. We have only to show that these are

† Then we say that $(x_1, \ldots x_n)$ is a *totally space-like point.*

points of holomorphy of both functions. This is achieved most simply by applying the edge of the wedge theorem. Consider the distribution

$$W' = (\Psi_0, \varphi_n(x_n)\ldots\varphi_2(x_2)\varphi_1(x_1)\Psi_0).$$

Then the corresponding holomorphic function W' is holomorphic if $-\eta_j \in V_+$, by the previous theorem. Furthermore, the boundary values W' and W agree at totally space-like points. Hence, by the edge of the wedge theorem W is holomorphic at such a point. ▮

The theorem also follows directly from the fact that the extended tube and the permuted extended tubes have a real environment in common (see Figure 2–4).

The expectation values for a theory of a free field can be computed directly from the definition of the field operators (3–14). For example, for a hermitian scalar field the state Ψ_0 is represented by $(1, 0, 0,\ldots)$, the state $\varphi(f)\Psi_0$ by $\sqrt{\pi}(0, \tilde{f}(p), 0,\ldots)$. The scalar product is then

$$(\Psi_0, \varphi(f)\,\varphi(g)\,\Psi_0) = \pi\int \tilde{\bar{f}}(p)\,\tilde{g}(p)\,d\Omega_m(p)$$
$$= 2\pi\int \tilde{\bar{f}}(p)\,\tilde{g}(p)\,\theta(p_0)\,\delta(p^2 - m^2)\,d^4p.$$

It follows that as a distribution

$$(\Psi_0, \varphi(x)\varphi(y)\Psi_0) = \frac{1}{i}\,\Delta^+(x - y; m),$$

where

$$\Delta^+(x; m) = \frac{i}{(2\pi)^3}\int \theta(p_0)\,\delta(p^2 - m^2)\,e^{-ipx}\,d^4p.$$

A repeated application of (3–14) gives

$$(\Psi_0, \varphi(x_1)\varphi(x_2)\cdots\varphi(x_n)\Psi_0)$$
$$= \begin{cases} \displaystyle\sum_{\text{partitions}} \frac{1}{i}\,\Delta^+(x_{i_1} - x_{i_2})\,\frac{1}{i}\,\Delta^+(x_{i_3} - x_{i_4}) \cdots \frac{1}{i}\,\Delta^+(x_{i_{n-1}} - x_{i_n}) & n \text{ even} \\ 0 & n \text{ odd} \end{cases} \tag{3–41}$$

where the sum is over all partitions of n into $n/2$ disjoint two element subsets $(i_1 i_2)(i_3 i_4)\ldots(i_{n-1} i_n)$ with $i_{2k-1} < i_{2k}$. We leave the proof to the reader. The study of the properties of the holomorphic functions W of a field theory has been called the *linear program*. An up-to-date summary of progress is to be found in Ref. 12. The idea is that once we have computed the domain of holomorphy explicitly, we can express the function in terms of its boundary values using a generalization of Cauchy's integral formula. It is hoped that the study of such integral representations is easier than the direct study of operator distributions satisfying local commutativity. A complete characterization of functions W with the properties given by the various theorems of this section is important, since, as the reconstruction theorem (Theorem 3–7) shows, such functions can be used to construct a field theory satisfying all the axioms except asymptotic completeness. The study of the last property leads to non-linear integral equations among the vacuum expectation values; this gives rise to the *non-linear program* (Ref. 16).

3–4. THE RECONSTRUCTION THEOREM: RECOVERY OF A THEORY FROM ITS VACUUM EXPECTATION VALUES

We give the complete proof of the reconstruction theorem only for the theory of a hermitian scalar field. For a general field theory, the theorem is true, but to write out the proof would require a massive automation of the notation.

Theorem 3–7

Let $\{\mathscr{W}^{(n)}\}$, $n = 1, 2,\ldots$ be a sequence of tempered distributions, where $\mathscr{W}^{(n)}$ depends on n four-vector variables $x_1, x_2,\ldots x_n$. Suppose the $\mathscr{W}^{(n)}$ have the cluster decomposition property of Theorem 3–4 and properties (a) through (e) of Theorems 3–1, 3–2, and 3–3, appropriately specialized to the case of a hermitian scalar field. That is,

$$\lim_{\lambda\to\infty} [\mathscr{W}(x_1,\ldots x_j, x_{j+1} + \lambda a,\ldots x_n + \lambda a) - \mathscr{W}(x_1,\ldots x_j)\mathscr{W}(x_{j+1},\ldots x_n)] = 0$$

if a is a space-like four-vector.

(*a*) (*Relativistic Transformation Law*)

$$\mathscr{W}^{(n)}(x_1,\ldots x_n) = \mathscr{W}^{(n)}(\Lambda x_1 + a, \Lambda x_2 + a,\ldots \Lambda x_n + a) \qquad \Lambda \in L_+^\uparrow. \tag{3–42}$$

(*b*) (*Spectral Conditions*)

$$\tilde{\mathscr{W}}^{(n)}(p_1,\ldots p_n) = (2\pi)^4\delta\left(\sum_{j=1}^n p_j\right)\tilde{W}(p_1, p_1 + p_2,\ldots p_1 + p_2 + \cdots + p_{n-1}) \tag{3–43}$$

and

$$\tilde{W}^{(n)}(q_1,\ldots q_{n-1}) = 0 \quad \text{if any } q_i \notin V_+. \tag{3–44}$$

(c) (*Hermiticity Conditions*)

$$\mathscr{W}^{(n)}(x_1, \ldots x_n) = \overline{\mathscr{W}^{(n)}(x_n, \ldots x_1)}. \qquad (3\text{–}45)$$

(d) (*Local Commutativity Conditions*)

$$\mathscr{W}^{(n)}(x_1, \ldots x_j, x_{j+1}, \ldots x_n) = \mathscr{W}^{(n)}(x_1, \ldots x_{j+1}, x_j, \ldots x_n)$$
$$\text{if } (x_j - x_{j+1})^2 < 0 \qquad \text{for } j = 1, 2, \ldots n-1. \qquad (3\text{–}46)$$

(e) (*Positive Definite Conditions*)

$$\sum \int \ldots \int dx_1 \ldots dx_n \, dy_1 \ldots dy_n \overline{f_j(x_1, \ldots x_j)}$$
$$\times \mathscr{W}^{(j+k)}(x_j, \ldots x_1, y_1, \ldots y_k) f_k(y_1, \ldots y_k) \geqslant 0 \quad (3\text{–}47)$$

for all finite sequences $f_0, f_1(x_1), f_2(x_1, x_2), \ldots$ of test functions.

Then there exists a separable† Hilbert space $\mathscr{H}$, a continuous unitary representation $U(a, \Lambda)$ of $\mathscr{P}_+^\uparrow$ in $\mathscr{H}$, a unique state Ψ_0, invariant under $U(a, \Lambda)$, and a hermitian scalar field with domain D_1 and representation of $\mathscr{P}_+^\uparrow$, $U(a, \Lambda)$, such that

$$(\Psi_0, \varphi(x_1) \ldots \varphi(x_n)\Psi_0) = \mathscr{W}^{(n)}(x_1, x_2, \ldots x_n).$$

Furthermore, any other field theory with these vacuum expectation values is unitary equivalent to this one. That is, if $\mathscr{H}_1$ is a Hilbert space, and $\{a, \Lambda\} \to U_1(a, \Lambda)$ is a continuous unitary representation of $\mathscr{P}_+^\uparrow$ in it, and Ψ_{01} is a unique vector in $\mathscr{H}_1$ invariant under $U_1(a, \Lambda)$, and $\varphi_1(x)$ is a scalar field with domain D_{11} with the property

$$(\Psi_{01}, \varphi_1(x_1) \ldots \varphi_1(x_n)\Psi_{01}) = \mathscr{W}^{(n)}(x_1, \ldots x_n)$$

and having Ψ_{01} as cyclic vector, then there exists a unitary transformation V of $\mathscr{H}$ onto $\mathscr{H}_1$ such that

$$\Psi_{01} = V\Psi_0, \qquad U_1(a, \Lambda) = VU(a, \Lambda)V^{-1},$$
$$\varphi_1(h) = V\varphi(h)V^{-1}, \qquad D_{11} = VD_1.$$

Proof of the Reconstruction Theorem:‡

To construct the Hilbert space $\mathscr{H}$ we begin with a vector space, H, formed of all sequences $f = (f_0, f_1, f_2, \ldots)$ where f_0 is any complex constant, $f_k \in \mathscr{S}(\mathbf{R}^{4k})$, $k = 1, 2, \ldots$, and all but a finite number of

† Defined in Section 2–6.

‡ In this proof h will denote a test function in $\mathscr{S}(\mathbf{R}^4)$ and f and g will denote elements of H, defined below.

test functions $f_k = 0$. Addition and multiplication by complex scalars are defined in the usual way:

$$(f_0, f_1, \ldots) + (g_0, g_1, \ldots) = (f_0 + g_0, f_1 + g_1, \ldots)$$
$$\alpha(f_0, f_1, f_2, \ldots) = (\alpha f_0, \alpha f_1, \alpha f_2, \ldots).$$

Scalar multiplication is commutative, associative, and distributive with respect to the addition.

Next we introduce a scalar product defined on pairs of vectors of the vector space

$$(f, g) = \sum_{j,k=0}^{\infty} \int \ldots \int dx_1 \ldots dx_j \, dy_1 \ldots dy_k$$
$$\times \overline{f_j(x_1, \ldots x_j)} \mathscr{W}^{(j+k)}(x_j, \ldots x_1, y_1, \ldots y_k) g_k(y_1, \ldots y_k), \quad (3\text{–}48)$$

where $\mathscr{W}^{(0)} \equiv 1$ by definition. This scalar product has the required property

$$(f, g) = \overline{(g, f)}$$

by virtue of the hermiticity condition (3–45) on the $\mathscr{W}$. Its linearity in g and anti-linearity in f are evident from (3–48). Furthermore, from the positive definiteness conditions (3–47) we get $\|f\|^2 = (f, f) \geqslant 0$. The term $\|f\|$ is called the norm of f. Now we define a linear transformation $U(a, \Lambda)$ of the vector space by

$$U(a, \Lambda)(f_0, f_1, f_2, \ldots) = (f_0, \{a, \Lambda\}f_1, \{a, \Lambda\}f_2, \ldots),$$

where,

$$\{a, \Lambda\}f_k(x_1, \ldots x_k) = f_k(\Lambda^{-1}(x_1 - a), \ldots \Lambda^{-1}(x_k - a)). \quad (3\text{–}49)$$

If we denote the vector $(1, 0, 0, \ldots)$ by Ψ_0 we have clearly

$$U(a, \Lambda)\Psi_0 = \Psi_0.$$

The operator $U(a, \Lambda)$ leaves the scalar product invariant by virtue of the relativistic transformation law (3–42) of the $\mathscr{W}$, and is a representation of $\mathscr{P}_+^\uparrow$, that is,

$$U(a_1, \Lambda_1)U(a_2, \Lambda_2) = U(a_1 + \Lambda_1 a_2, \Lambda_1 \Lambda_2).$$

This is proved easily using (3–49).

We introduce a linear operator $\varphi(h)$ for each test function h by the equation

$$\varphi(h)\{f_0, f_1, f_2, \ldots\} = (0, hf_0, h \otimes f_1, h \otimes f_2, \ldots), \quad (3\text{–}50)$$

where

$$(h \otimes f_k)(x_1, \ldots x_{k+1}) = h(x_1) f_k(x_2, x_3, \ldots x_{k+1})$$

is clearly a test function. That $\varphi(h)$ satisfies the transformation law

$$U(a, \Lambda)\varphi(h)U(a, \Lambda)^{-1} = \varphi(\{a, \Lambda\}h) \quad (3\text{–}51)$$

follows immediately from the computation

$$\begin{aligned} U(a, \Lambda)\varphi(h)(f_0, f_1, \ldots) &= U(a, \Lambda)(0, hf_0, h \otimes f_1, \ldots) \\ &= (0, \{a, \Lambda\}hf_0, \{a, \Lambda\}h \otimes \{a, \Lambda\}f_1, \ldots) \\ &= \varphi(\{a, \Lambda\}h)(f_0, \{a, \Lambda\}f_1, \ldots) \\ &= \varphi(\{a, \Lambda\}h)U(a, \Lambda)(f_0, f_1, \ldots). \end{aligned}$$

As functionals of h the matrix elements $(f, \varphi(h)g)$ are tempered distributions since they are finite sums of $\mathscr{W}$'s, and further

$$(\varphi(\bar{h})f, g) = (f, \varphi(h)g). \tag{3–52}$$

These constructions have given us a Ψ_0, $U(a, \Lambda)$ and $\varphi(h)$, and a vector space H, but H need not be a Hilbert space. It can be deficient in two respects. First, it may contain vectors of zero norm in the scalar product (3–48). Second, it may not be complete.† We will remedy these two deficiencies in turn. The reader who is familiar with the standard mathematical operation of completing a pre-Hilbert space to get a Hilbert space can skip most of the following few paragraphs, which are chiefly an outline of the required constructions.

First, note that the set $H_0 \subset H$ of all vectors of zero norm forms an isotropic subspace, that is, a subspace in which each vector is orthogonal to every other vector. To see this, note that if $\|f\| = 0$, then for any g,

$$|(f, g)| \leqslant \|f\| \, \|g\| = 0 \tag{3–53}$$

by the Schwarz inequality (which is valid as long as the scalar product is non-negative). Thus, if $f = (f_0, f_1, \ldots)$ and $g = (g_0, g_1, \ldots)$ are of zero norm, then f is orthogonal to g and $\alpha f + \beta g$ is of zero norm. We now form equivalence classes of sequences, $f = (f_0, f_1, \ldots)$, two sequences being equivalent if they differ by a sequence of zero norm. These equivalence classes form in a natural way a vector space, usually denoted by H/H_0, on which the scalar product is positive definite. If $\mathfrak{f}$ and $\mathfrak{g}$ are two equivalence classes, we merely define $\alpha\mathfrak{f} + \beta\mathfrak{g}$ to be that equivalence class to which $\alpha f + \beta g$ belongs, where $f = (f_0, f_1, \ldots)$ belongs to $\mathfrak{f}$ and $g = (g_0, g_1, \ldots)$ to $\mathfrak{g}$. The result is independent of which representatives are chosen because the set of vectors of zero length is a vector space. Clearly, the set of sequences of zero norm serves as the zero in the vector space H/H_0 of equivalence classes, and $\|\mathfrak{f}\| = 0$ implies $\mathfrak{f} = 0$. We can define a scalar product in H/H_0 by $(\mathfrak{f}, \mathfrak{g}) = (f, g)$; it is independent of which representative $f \in \mathfrak{f}$ is chosen by virtue of (3–53).

We have next to verify that the $U(a, \Lambda)$ and $\varphi(h)$ defined by (3–49) and (3–50) are actually mappings of equivalence classes. For $U(a, \Lambda)$

† Defined in Section 2–6.

this is an immediate consequence of the fact that it leaves scalar products invariant: if f and g are two sequences of test functions for which

$$\|f - g\| = 0$$

then

$$\|U(a, \Lambda)f - U(a, \Lambda)g\| = \|U(a, \Lambda)(f - g)\| = \|f - g\| = 0,$$

that is, if f and g are in the same equivalence class, so are $U(a, \Lambda)f$ and $U(a, \Lambda)g$. That $\|f\| = 0$ implies $\|\varphi(h)f\| = 0$ follows from (3–52) and the Schwarz inequality (3–53):

$$\begin{aligned}(\varphi(h)f, \varphi(h)f) &= (f, \varphi(\bar{h})\varphi(h)f)\\ &\leqslant \|f\|\ \|\varphi(\bar{h})\varphi(h)f\| = 0 \qquad \text{if } \|f\| = 0.\end{aligned}$$

For brevity we shall use the same symbols $U(a, \Lambda)$, $\varphi(h)$ to denote operators acting on the equivalence classes (that is, operators in H/H_0) as we used for the original operators in H. Similarly $\Psi_0 \in H/H_0$ will denote the equivalence class of $(1, 0, 0, \ldots)$.

Next we have to deal with the incompleteness of the space H/H_0 of equivalence classes. This is a standard problem analogous to completing the rational numbers to get the real numbers. Consider the space $\mathfrak{h}$ of all Cauchy sequences $F = \{\mathfrak{f}_1, \mathfrak{f}_2, \ldots\}$ of elements $\mathfrak{f}_j \in H/H_0$, that is, sequences such that

$$\|\mathfrak{f}_m - \mathfrak{f}_n\| \to 0$$

as m and $n \to \infty$. $\mathfrak{h}$ is a vector space with addition defined by

$$\alpha\{\mathfrak{f}_1, \mathfrak{f}_2, \ldots\} + \beta\{\mathfrak{g}_1, \mathfrak{g}_2, \ldots\} = \{\alpha\mathfrak{f}_1 + \beta\mathfrak{g}_1, \alpha\mathfrak{f}_2 + \beta\mathfrak{g}_2, \ldots\}.$$

The scalar product of two elements $F = \{\mathfrak{f}_1, \mathfrak{f}_2, \ldots\}$ and $G = \{\mathfrak{g}_1, \mathfrak{g}_2, \ldots\}$ is defined by

$$(F, G) = \lim_{n\to\infty} (\mathfrak{f}_n, \mathfrak{g}_n). \tag{3–54}$$

The existence of this limit is an immediate consequence of the assumption that F and G are Cauchy sequences [*Proof:* because $\{\mathfrak{f}_1, \mathfrak{f}_2, \ldots\}$ is a Cauchy sequence, $\|\mathfrak{f}_n\|$ must be bounded. Consequently

$$\begin{aligned}|(\mathfrak{f}_n, \mathfrak{g}_n) - (\mathfrak{f}_m, \mathfrak{g}_m)| &= |(\mathfrak{f}_n - \mathfrak{f}_m, \mathfrak{g}_n) + (\mathfrak{f}_m, \mathfrak{g}_n - \mathfrak{g}_m)|\\ &\leqslant \|\mathfrak{g}_n\|\ \|\mathfrak{f}_n - \mathfrak{f}_m\| + \|\mathfrak{f}_m\|\ \|\mathfrak{g}_n - \mathfrak{g}_m\|,\end{aligned}$$

which tends to zero for large m and n. The convergence of (3–54) follows.] Elementary calculations which are left to the reader show that (F, G) defines a scalar product. However, just as in the case of

H, there may be vectors of zero norm in $\mathfrak{h}$. These again form an isotropic subspace $\mathfrak{h}_0$, and again we have to introduce equivalence classes of vectors, two vectors being equivalent if they differ by a vector of zero norm. The space of these equivalence classes we denote by $\mathscr{H} = \mathfrak{h}/\mathfrak{h}_0$. It is a vector space with positive definite scalar product. Furthermore, it is complete. This is proved as follows. Let $\Phi_1, \Phi_2, \ldots$ be a Cauchy sequence of elements of $\mathscr{H}$. This means $\|\Phi_m - \Phi_n\|$ is arbitrarily small for all sufficiently large m and n. Therefore, any representatives of the equivalence classes $\Phi_1, \Phi_2, \ldots$, say $F_1, F_2, \ldots$ respectively, where $F_j \in \mathfrak{h}$, satisfy

$$\|F_n - F_m\| = \|\Phi_n - \Phi_m\| \to 0 \qquad \text{as } n, m \to \infty.$$

If $F_j = \{\mathfrak{f}_1^{(j)}, \mathfrak{f}_2^{(j)}, \ldots\}$, then, by definition,

$$\|F_n - F_m\| = \lim_{k \to \infty} \|\mathfrak{f}_k^{(n)} - \mathfrak{f}_k^{(m)}\|.$$

We may assume that the representative sequences F_n are chosen so that $\|\mathfrak{f}_k^{(n)} - \mathfrak{f}_k^{(m)}\| < 2\|\Phi_n - \Phi_m\|$ and $\|\mathfrak{f}_i^{(k)} - \mathfrak{f}_j^{(k)}\| < 1/k$ for all i, j, k and all m and n. This can always be arranged by dropping enough $\mathfrak{f}_k^{(n)}$, at the beginning of the sequences F_n. We now show that $G = \{\mathfrak{f}_1^{(1)}, \mathfrak{f}_2^{(2)}, \ldots\}$ is a Cauchy sequence such that

$$\lim_{n \to \infty} \|F_n - G\| = 0.$$

The equivalence class of G, which we shall call Φ, is then an element of $\mathscr{H}$ such that $\lim_{n\to\infty} \|\Phi_n - \Phi\| = 0$, proving that $\mathscr{H}$ is complete. To prove our assertions requires a few more lines. We have

$$\begin{aligned}\|\mathfrak{f}_k^{(k)} - \mathfrak{f}_\ell^{(\ell)}\| &\leqslant \|\mathfrak{f}_k^{(k)} - \mathfrak{f}_\ell^{(k)}\| + \|\mathfrak{f}_\ell^{(k)} - \mathfrak{f}_\ell^{(\ell)}\| \\ &\leqslant \|\mathfrak{f}_k^{(k)} - \mathfrak{f}_\ell^{(k)}\| + 2\|\Phi_k - \Phi_l\| \to 0 \quad \text{as } k, \ell \to \infty.\end{aligned}$$

This proves that $\{\mathfrak{f}_k^{(k)}\}$, $k = 1, 2, \ldots$ is a Cauchy sequence. Again,

$$\|F_n - G\| = \lim_{k \to \infty} \|\mathfrak{f}_k^{(n)} - \mathfrak{f}_k^{(k)}\|$$

and each term on the right-hand side is less than $2\|\Phi_n - \Phi_k\|$, which goes to zero as $n, k \to \infty$. Therefore $\|F_n - G\| \to 0$.

This completes the construction of our Hilbert space of states. When a theory is given in terms of its vacuum expectation values, a state vector $\Psi \in \mathscr{H} = \mathfrak{h}/\mathfrak{h}_0$ is an equivalence class (modulo the equivalence class $\mathfrak{h}_0$ of $\{0, 0, \ldots\}$) of Cauchy sequences of equivalence classes $\in H/H_0$ (modulo the equivalence class H_0 of sequences of zero norm) of sequences of test functions. Yes, it really is that simple!

The elements of H/H_0 can be thought of as embedded in $\mathscr{H}$. For each element, $\mathfrak{f}$ of H/H_0 one has a vector $\Psi_{\mathfrak{f}}$ in $\mathscr{H}$, defined as the equivalence class of the Cauchy sequence $F = \{\mathfrak{f}, \mathfrak{f}, \mathfrak{f}, \ldots\} \in \mathfrak{h}$. Clearly, the scalar product $(\Psi_{\mathfrak{f}}, \Psi_{\mathfrak{g}})$ in $\mathscr{H}$ is equal to $(\mathfrak{f}, \mathfrak{g})$, the scalar product in H/H_0. The $\Psi_{\mathfrak{f}}$ constitute a subset D_1 of $\mathscr{H}$ which is dense in $\mathscr{H}$, for if Φ is an element of $\mathscr{H}$ and $\{\mathfrak{f}_1, \mathfrak{f}_2, \ldots\}$ is a representative Cauchy sequence belonging to Φ, $\|\mathfrak{f}_n - \mathfrak{f}_m\| < \varepsilon$ for all $m, n > N$ implies $\|\Phi - \Psi_{\mathfrak{f}_n}\| = \lim_{m\to\infty} \|\mathfrak{f}_m - \mathfrak{f}_n\| \leqslant \varepsilon$.

We can use the fact that the $\Psi_{\mathfrak{f}}$ are dense in $\mathscr{H}$ to show that $\mathscr{H}$ is separable. The required denumerable dense set is obtained by taking sequences $\{f_0, f_1, \ldots\} \in H$, where the f_j are test functions chosen from the denumerable set dense in $\mathscr{S}(\mathbf{R}^{4j})$ described in Section 2–1. We leave it to the reader to verify that this choice of $\Psi_{\mathfrak{f}}$ is dense in $\mathscr{H}$.

Up to this point $U(a, \Lambda)$ has only been defined on H/H_0. This immediately yields a definition of it on the dense set of vectors $\Psi_{\mathfrak{f}}$ in $\mathscr{H}$. Since it is continuous there, because

$$\|U(a, \Lambda)\Psi_{\mathfrak{f}} - U(a, \Lambda)\Psi_{\mathfrak{g}}\| = \|\Psi_{\mathfrak{f}} - \Psi_{\mathfrak{g}}\|,$$

it can be extended by continuity to all of $\mathscr{H}$; and so extended, is by definition scalar product-preserving: if $\Psi_{\mathfrak{f}_n} \to \Phi$ and $\Psi_{\mathfrak{g}_n} \to \chi$ then, by definition, $U(a, \Lambda)\Psi_{\mathfrak{f}_n} \to U(a, \Lambda)\Phi$ and $U(a, \Lambda)\Psi_{\mathfrak{g}_n} \to U(a, \Lambda)\chi$, so

$$(U(a, \Lambda)\Phi, U(a, \Lambda)\chi) = \lim_{n,m\to\infty} (U(a, \Lambda)\Psi_{\mathfrak{f}_n}, U(a, \Lambda)\Psi_{\mathfrak{g}_m}) = (\Phi, \Psi).$$

Furthermore, so defined on $\mathscr{H}$, $U(a, \Lambda)$ is continuous in $\{a, \Lambda\}$. Because $\|(U(b, M)\Psi - U(a, \Lambda)\Psi\| = \|\Psi - U(\{b, M\}^{-1}\{a, \Lambda\})\Psi\|$ it is sufficient to verify the continuity at the identity element $\{0,1\}$. On vectors of the form $\Psi_{\mathfrak{f}}$ this continuity is easy to check:

$$\|\Psi_{\mathfrak{f}} - U(a, \Lambda)\Psi_{\mathfrak{f}}\| = \|\mathfrak{f} - U(a, \Lambda)\mathfrak{f}\|. \tag{3–55}$$

Since the right-hand side can be expressed in terms of the distributions $\mathscr{W}$ smeared with test functions f_k and $\{a, \Lambda\}f_\ell$, and $\{a, \Lambda\}f_\ell$ depends continuously on $\{a, \Lambda\}$ in the topology of $\mathscr{S}(\mathbf{R}^{4\ell})$, we see that (3–55) is continuous in $\{a, \Lambda\}$. The continuity of $U(a, \Lambda)$ on a general $\Psi \in \mathscr{H}$ then follows from the fact that it is unitary. Explicitly

$$\begin{aligned}\|\Psi - U(a, \Lambda)\Psi\| &= \|(\Psi - \Psi_{\mathfrak{f}}) + (\Psi_{\mathfrak{f}} - U(a, \Lambda)\Psi_{\mathfrak{f}}) \\ &\qquad + U(a, \Lambda)(\Psi_{\mathfrak{f}} - \Psi)\| \\ &\leqslant 2\|\Psi - \Psi_{\mathfrak{f}}\| + \|\Psi_{\mathfrak{f}} - U(a, \Lambda)\Psi_{\mathfrak{f}}\| \to 0 \\ &\qquad \text{as } (a, \Lambda) \to (0, 1) \text{ and } \Psi_{\mathfrak{f}} \to \Psi.\end{aligned}$$

The $U(a, \Lambda)$ possess an invariant vector, which with a final change in notation we again denote Ψ_0. It is the equivalence class of the Cauchy sequence $\{\mathfrak{f}, \mathfrak{f}, \ldots\}$ of $\mathfrak{h}$, where $\mathfrak{f} \in H/H_0$ is the equivalence class of the sequence of test functions $(1, 0, 0, \ldots)$. We now show that by virtue of the cluster decomposition property there is no other state Ψ_0' linearly independent of Ψ_0 and invariant under $U(a, \Lambda)$. Without loss of generality we can assume $(\Psi_0', \Psi_0) = 0$, and $(\Psi_0', \Psi_0') = 1$. If it happened that Ψ_0' were of the form $\Psi_{\mathfrak{f}} \in D_1$ we would have an immediate contradiction because, for a space-like vector a,

$$(\Psi_0', \Psi_0') = \lim_{\lambda \to \infty} (\Psi_0', U(\lambda a, 1)\Psi_0') = (\Psi_0', \Psi_0)(\Psi_0, \Psi_0') = 0,$$

where the second equality follows from the cluster decomposition property of the vacuum expectation values:

$$\begin{aligned}(\Psi_{\mathfrak{f}}, U(\lambda a, 1)\Psi_{\mathfrak{f}}) &= ((f_0, f_1, \ldots), (f_0, \{\lambda a, 1\}f_1, \{\lambda a, 1\}f_2, \ldots)) \\ &= \sum_{j,k=0}^{\infty} \int \cdots \int dx_1 \ldots dx_j\, dy_1 \ldots dy_k \bar{f}_j(x_1, \ldots x_j) \\ &\qquad \times \mathscr{W}^{(j+k)}(x_j, \ldots x_1, y_1 + \lambda a, \ldots y_k + \lambda a) f_k(y_1, \ldots y_k) \\ &\to \sum_{j,k=0}^{\infty} \int \cdots \int dx_1 \ldots dx_j\, dy_1 \ldots dy_k \bar{f}_j(x_1, \ldots x_j) \mathscr{W}^{(j)}(x_j, \ldots x_1) \\ &\qquad \times \mathscr{W}^{(k)}(y_1, \ldots y_k) f_k(y_1, \ldots y_k) \\ &= (\Psi_{\mathfrak{f}}, \Psi_0)(\Psi_0, \Psi_{\mathfrak{f}}).\end{aligned}$$

In general, Ψ_0' is not of the form $\Psi_{\mathfrak{f}}$ but because the $\Psi_{\mathfrak{f}}$ are dense, it can be approximated by $\Psi_{\mathfrak{f}}$'s to arbitrary accuracy. Suppose $\|\Psi_0' - \Psi_{\mathfrak{f}}\| < \varepsilon$, where without loss of generality we may take $\|\Psi_{\mathfrak{f}}\| = 1$. Then

$$\begin{aligned}(\Psi_0', \Psi_0') &= \lim_{\lambda \to \infty} (\Psi_0', U(\lambda a, 1)\Psi_0') \\ &= \lim_{\lambda \to \infty} [(\Psi_0' - \Psi_{\mathfrak{f}}, U(\lambda a, 1)\Psi_0') \\ &\qquad + (\Psi_{\mathfrak{f}}, U(\lambda a, 1)(\Psi_0' - \Psi_{\mathfrak{f}})) + (\Psi_{\mathfrak{f}}, U(\lambda a, 1)\Psi_{\mathfrak{f}})]\end{aligned}$$

and therefore, by the cluster decomposition theorem,

$$|(\Psi_0', \Psi_0') - (\Psi_{\mathfrak{f}}, \Psi_0)(\Psi_0, \Psi_{\mathfrak{f}})| \leqslant 2\varepsilon.$$

But

$$|(\Psi_{\mathfrak{f}}, \Psi_0)| = |(\Psi_{\mathfrak{f}} - \Psi_0', \Psi_0)| \leqslant \|\Psi_{\mathfrak{f}} - \Psi_0'\| < \varepsilon$$

so

$$\begin{aligned}(\Psi_0', \Psi_0') &\leqslant (\Psi_{\mathfrak{f}}, \Psi_0)(\Psi_0, \Psi_{\mathfrak{f}}) + |(\Psi_0', \Psi_0') - (\Psi_{\mathfrak{f}}, \Psi_0)(\Psi_0, \Psi_{\mathfrak{f}})| \\ &\leqslant 2\varepsilon + \varepsilon^2,\end{aligned}$$

which is a contradiction for sufficiently small ε. Thus, the vacuum state is unique.

There remains only one property of $U(a, \Lambda)$ to be verified, namely, that its energy momentum spectrum lies in or on the future light cone. This is an immediate consequence of (3–44).

The smeared field $\varphi(h)$ was defined on all vectors $\mathfrak{g}$ of H/H_0. This leads at once to a definition on the subset D_1 of vectors $\Psi_\mathfrak{g}$ of $\mathscr{H}$. We define $\varphi(h)\Psi_\mathfrak{g}$ to be $\Psi_{h\otimes\mathfrak{g}}$, where $h \otimes \mathfrak{g}$ is the equivalence class of the sequence

$$(0, hg_0, h \otimes g_1, h \otimes g_2, \ldots) \tag{3–56}$$

$(g_0, g_1, \ldots)$ being a member of the equivalence class $\mathfrak{g}$; with this definition

$$\varphi(h)[\alpha\Psi_\mathfrak{g} + \beta\Psi_\mathfrak{f}] = \Psi_{\alpha h\otimes\mathfrak{g} + \beta h\otimes\mathfrak{f}}.$$

In particular this defines $\varphi(h)$ on D_0, the set of vectors of the form $\mathscr{P}[\varphi(h)]\Psi_0$. Furthermore,

$$\int\cdots\int f_1(x_1)\cdots f_n(x_n)\, dx_1\cdots dx_n \mathscr{W}^{(n)}(x_1,\ldots x_n) = (\Psi_0, \varphi(f_1)\ldots\varphi(f_n)\Psi_0).$$

The domain D_1 is suitable as a domain of definition of a field, for it clearly satisfies $\varphi(h)D_1 \subset D_1$, $\Psi_0 \in D_1$, $U(a, \Lambda)D_1 \subset D_1$. Since $(\Psi_\mathfrak{g}, \varphi(h)\Psi_\mathfrak{f})$ is a finite linear combination of $\mathscr{W}$'s it is a tempered distribution in h. This completes the verification of axioms O and I. The transformation law II is an immediate consequence of (3–42). The hermiticity relation $\varphi(h)\Psi = [\varphi(\bar{h})]^*\Psi$ can be proved for all vectors Ψ in D_1. Since φ and φ^* are linear operators it is sufficient to verify the equation for vectors of the form

$$\Psi_\mathfrak{g} = \int g(x_1,\ldots x_n)\varphi(x_1)\ldots\varphi(x_n)\, dx_1\ldots dx_n\, \Psi_0.$$

It is a simple consequence of (3–56) that

$$\|\varphi(h)\Psi_\mathfrak{g} - \varphi(\bar{h})^*\Psi_\mathfrak{g}\| = 0$$

using the definition of the norm in terms of the $\mathscr{W}$ given by $\|\Psi_\mathfrak{g}\|^2 = (\Psi_\mathfrak{g}, \Psi_\mathfrak{g})$. In the same way one verifies the local commutativity condition for vectors in D_1 using (3–56).

This completes the explicit construction of a hermitian scalar field. It remains to show that any other field with the same vacuum expectation values is unitary equivalent to the one so constructed.

Suppose that $\mathscr{H}_1$, $U_1(a, \Lambda)$, and $\varphi_1(x)$ define a theory with vacuum Ψ_{01} having the same expectation values. Let V be the mapping taking $\Psi_\mathfrak{f} \in \mathscr{H}$ to a vector in $\mathscr{H}_1$ given by

$$V\Psi_\mathfrak{f} = \Psi_{1\mathfrak{f}} = f_0\Psi_{01} + \varphi_1(f_1)\Psi_{01} + \int\varphi_1(x_1)\varphi_1(x_2)f_2(x_1, x_2)\, dx_1\, dx_2\, \Psi_{01} + \cdots, \tag{3–57}$$

where $\{f_0, f_1, f_2, \ldots\} = f$ lies in the equivalence class $\mathfrak{f}$. The mapping

V is well defined because (3–57) is obviously independent of which representative is chosen. The mapping V is unitary because

$$(V\Psi_{\mathfrak{f}}, V\Psi_{\mathfrak{g}}) = (\Psi_{1\mathfrak{f}}, \Psi_{1\mathfrak{g}})$$

follows from the equality of the expectation values. This can be extended by continuity to the whole of $\mathscr{H}$ and $\mathscr{H}_1$ [we assume vectors of the form (3–57) are dense in $\mathscr{H}_1$]. If $\Psi_{\mathfrak{f}} \in D_1$ then

$$V\varphi(h)\Psi_{\mathfrak{f}} = V\Psi_{h\otimes\mathfrak{f}},$$

where $\Psi_{h\otimes\mathfrak{f}}$ is defined by (3–57)

$$\begin{aligned} &= \varphi_1(hf_0)\Psi_{01} + \int h(x_1)f_1(x_2)\varphi_1(x_1)\varphi_1(x_2)\, dx_1\, dx_2\, \Psi_{01} + \cdots \\ &= \varphi_1(h)[f_0\Psi_{01} + \int f(x_1)\varphi(x_1)\, dx_1\, \Psi_{01} + \cdots] \\ &= \varphi_1(h)V\Psi_{\mathfrak{f}}. \end{aligned}$$

Hence $V\varphi(h)V^{-1} = \varphi_1(h)$ holds for all vectors in $D_{11} \subset \mathscr{H}_1$. Finally, a simple direct calculation shows that $U_1(a, \Lambda) = VU(a, \Lambda)V^{-1}$. ∎

We remark that this reconstruction can be carried through for a denumerable set of fields of any spin, provided of course we are given enough $\mathscr{W}$'s with the correct properties. We leave it to the reader to carry through the construction for the free field and generalized free field, starting from the vacuum expectation values given in Section 3–3. The reconstruction will not (from the listed properties alone) give a theory satisfying asymptotic completeness. However, if the spectrum of the theory shows an isolated representation of $\mathscr{P}_+^\uparrow$ at $p^2 = m^2$, then the Haag–Ruelle theory does ensure a particle interpretation, at least for the collision states. We complete this chapter with a discussion of some other symmetries which might be present in the theory.

3–5. SYMMETRIES IN A FIELD THEORY

We now return to the point made at the end of Section 1–2, namely, that if a unitary operator in $\mathscr{H}$ is to be interpreted as a symmetry it must have a reasonable physical meaning. We first show that the assumption of a definite transformation law for all the fields of a field theory uniquely fixes the corresponding transformation on the states up to a phase.

For simplicity we prove the theorem only for a hermitian scalar field which transforms under the parity operator in the usual way. Simple alterations in the notation establish similar results in a general field theory for the operators P, C, and T given by (1–52).

Theorem 3–8

Let a field theory of a hermitian scalar field φ with domain of definition D be given. Suppose $U(I_s)$ is a unitary operator which satisfies

$$U(I_s)D = D$$
$$U(I_s)\varphi(x)U(I_s)^{-1} = \varphi(x_0, -\mathbf{x}) \equiv \varphi(I_s x). \qquad (3\text{–}58)$$

Then $U(I_s)$ is determined up to a phase, and $U\Psi_0 = e^{i\alpha}\Psi_0$. Further,

$$U(I_s)U(a, \Lambda)U(I_s)^{-1} = U(I_s a, I_s^{-1}\Lambda I_s), \qquad (3\text{–}59)$$

where $U(a, \Lambda)$ is the representation of $\mathscr{P}_+^\uparrow$.

Proof:

Choose Ψ, $\Phi \in D$ and consider

$$(\Psi, \varphi(x)U(I_s)U(a, \Lambda)U(I_s)^{-1}U(I_s a, I_s^{-1}\Lambda I_s)^{-1}\Phi).$$

By using (3–58) and (3–4), moving φ to the right, we transform this expression to

$$(\Psi, U(I_s)U(a, \Lambda)U(I_s)^{-1}U(I_s a, I_s^{-1}\Lambda I_s)^{-1}\varphi(x)\Phi).$$

Since, anticipating Theorem 4–5, we may assume φ is irreducible,

$$U(I_s)U(a, \Lambda)U(I_s)^{-1}U(I_s a, I_s^{-1}\Lambda I_s)^{-1} = \omega 1, \qquad (3\text{–}60)$$

where $|\omega| = 1$. Applying both sides of (3–60) to the vacuum and using $U(a, \Lambda)\Psi_0 = \Psi_0$ we see that

$$U(I_s a, I_s^{-1}\Lambda I_s)U(I_s)\Psi_0 = \omega^{-1}U(I_s)\Psi_0 \qquad (3\text{–}61)$$

for all a, Λ. Since the vacuum Ψ_0 is the only state invariant under all $U(a, \Lambda)$,

$$U(I_s)\Psi_0 = e^{i\alpha}\Psi_0.$$

Substituting in (3–61) we see $\omega = 1$, proving (3–59).

To show finally that $U(I_s)$ is determined up to a phase, suppose $U(I_s)\Psi_0 = \Psi_0$. Then, for states of the form $\Psi = \varphi(f_1)\ldots\varphi(f_n)\Psi_0$, we get

$$U(I_s)\Psi = U\varphi(f_1)U^{-1}U\varphi(f_2)U^{-1}\ldots U\varphi(f_n)U^{-1}U\Psi_0$$
$$= \varphi(\hat{f}_1)\ldots\varphi(\hat{f}_n)\Psi_0 \in D,$$

where $\hat{f}_i(x) = f_i(I_s x)$. Thus the operator $U(I_s)$ can be extended uniquely by linearity and continuity to the whole of $\mathscr{H}$, and, so extended, continues to satisfy (3–59). ▮

In the same way the requirements (1–52) for $U(C)$ and $U(I_t)$ lead to the properties $U(C)\Psi_0 = e^{i\alpha_C}\Psi_0$, $U(I_t)\Psi_0 = e^{i\alpha_t}\Psi_0$, and the "group" properties

$$U(I_t)U(a, \Lambda)U(I_t)^{-1} = U(I_t a, I_t^{-1}\Lambda I_t) \tag{3–62}$$

$$U(C)U(a, \Lambda)U(C)^{-1} = U(a, \Lambda). \tag{3–63}$$

Equations (3–59), (3–62), and (3–63) and the invariance of Ψ_0 are desirable properties of the operators $U(I_s)$, $U(I_t)$, and $U(C)$ for them to be interpreted as P, T, and C, respectively.

It is a standard convention to choose the arbitrary phase factor in the definition of $U(I_s)$, $U(I_t)$, or $U(C)$ so that

$$U(I_s)\Psi_0 = \Psi_0, \qquad U(I_t)\Psi_0 = \Psi_0, \qquad U(C)\Psi_0 = \Psi_0. \tag{3–64}$$

Then, the results of Theorem 3–8 can be summarized: *the requirement that the fields of a field theory have definite transformation laws under $U(I_s)$, $U(I_t)$, or $U(C)$ uniquely fixes those operators.*

Thus, in a field theory the problem of determining whether a given symmetry is a reasonable expression of the operations P, T, or C is reduced to the problem of understanding the physical content of transformation laws of the fields under those operations. This is a somewhat complicated subject whose full ramifications cannot be completely explored here. We shall content ourselves with a series of remarks.

For an observable field a transformation law under P, C, or T is a direct statement about observable quantities, and no possibility of ambiguity arises. For example, in the theory of neutral π-mesons the transformation law $\varphi(x) \to -\varphi(I_s x)$ appropriate to pseudo-scalar π^0-mesons is physically distinct from $\varphi(x) \to \varphi(I_s x)$ which would describe scalar π^0's. For unobservable fields on the other hand, the physical consequences of the assumption of a given transformation law are only indirectly ascertainable; an analysis of the super-selection rules of the theory is in general indispensable for deciding whether the assumption of a different transformation law actually leads to a physically distinct theory. For example, the choices of phase in the definitions (1–45) of P, C, and T for a single spin $\frac{1}{2}$ field ψ are rather arbitrary, as is the generalization to higher spin; the choice $\psi(x) \to \gamma^0\psi(I_s x)$ is just as good a definition of the parity operation as $\psi(x) \to -\gamma^0\psi(I_s x)$. If $U(I_s)$ is the transformation which produces the first choice, then $RU(I_s)$ will produce the second, if R is the operator which rotates through an angle 2π around some axis, i.e., R is 1 on the states of integer spin and -1 on the states of half-odd integer spin. In general, two different choices, U and U', for a symmetry operator are physically identical if there exists an operator V such that $U = VU'$ and $|(V\Psi, \Phi)|^2 = |(\Psi, \Phi)|^2$ for all physically realizable states Φ, Ψ.

A simple example in which the phases of the transformation law under $U(I_s)$ do have physical consequences is the case of two anti-commuting spin one-half fields, ψ_1 and ψ_2. Suppose $\psi_1(x) \to \varepsilon_1\gamma^0\psi_1(I_sx)$, $\psi_2(x) \to \varepsilon_2\gamma^0\psi_2(I_sx)$ under the operation of space-inversion. The transformation R described before can be used to obtain new fields ψ_1', ψ_2' which have $(\varepsilon_1, \varepsilon_2)$ replaced by $(-\varepsilon_1, -\varepsilon_2)$, but the theories with $(\varepsilon_1, \varepsilon_2) = (1, +1), (1, -1), (1, i)$ are physically distinct. For example, the quantity $\psi_1{}^+(x)\psi_2(y) + \psi_2{}^+(y)\psi_1(x)$ is a candidate for an observable in the first two cases, being scalar in the first and pseudo-scalar in the second, but in the third, it has no well-defined transformation law. The third case is interesting because in it $U(I_s)^2$ commutes with $\psi_2(x)$ and anti-commutes with $\psi_1(x)$. This implies the existence of a super-selection rule separating states of the form $P(\psi_1(f), \psi_2(g)\ldots)\Psi_0$, with P a polynomial odd in ψ_1, which span a subspace $\mathscr{H}_1$, from those in $\mathscr{H}_2$ which is spanned by states with P even in ψ_1. To see this recall that a physically realizable state $\boldsymbol{\Psi}$ must be unchanged when twice transformed by the parity operator, so if Ψ is a vector in a ray $\boldsymbol{\Psi}$, so is $U(I_s)^2\Psi$. Now $U(I_s)^2$ is -1 on $\mathscr{H}_1$ and $+1$ on $\mathscr{H}_2$, so a state represented by a vector $\alpha\Psi_1 + \beta\Psi_2$ with $\alpha\beta \neq 0$, $\Psi_1 \in \mathscr{H}_1$, and $\Psi_2 \in \mathscr{H}_2$ cannot be physically realizable. More generally, if $U(I_s)$, $U(C)$, or $U(I_t)$ are to be interpreted as P, C, or T, then each of $U(I_s)^2$, $U(I_t)^2$, $U(C)^2$, $U(I_s)U(C)U(I_s)U(C)$, etc., must be a constant multiple of the identity on every coherent subspace. (See Section 1–1 for the definition of physically realizable states and coherent subspaces.) What this result means in practice is that for certain choices of phases in $U(I_s)$, $U(I_t)$, or $U(C)$ theories are more general in the sense that they do not have super-selection rules forced on them by their invariance under P, C, or T. It was this kind of special choice which was made in Section 1–3.

A common situation in which different phases in the definition of inversions are physically equivalent is that in which a theory possesses a *multiplicative symmetry*. That means that there exists a unitary operator V such that for some set of complex numbers of absolute value 1

$$V\psi_j(x)V^{-1} = \lambda_j\psi_j(x). \tag{3–65}$$

The R described previously is an example of a multiplicative symmetry; in that case $\lambda_j = +1$ for integer spin fields and -1 for half-odd integer spin fields. By an argument just like that used in the proof of Theorem

3–8 one concludes that V commutes with $U(a, A)$ and leaves the vacuum state invariant. Thus

$$(\Psi_0, \psi_1(x_1) \ldots \Psi_0) = (V\Psi_0, V\psi_1(x_1) \ldots \Psi_0)$$
$$= \lambda_1^{k_1}\lambda_2^{k_2} \ldots (\Psi_0, \psi_1(x_1) \ldots \Psi_0),$$

where k_j is the number of times ψ_j appears in the vacuum expectation value. Clearly, $\lambda_1^{k_1}\lambda_2^{k_2} \ldots \neq 1$ implies that the corresponding vacuum expectation value is zero. The vanishing of such vacuum expectation values means that the theory possesses special selection rules. In such a theory the choice of $U(I_s)$ to represent space inversion can be physically equivalent to the choice $VU(I_s)$.

There is still another aspect of the transformation law of fields under symmetries which is essential if their physical interpretation is to be complete; that is the connection with collision theory. A straightforward application of the Haag–Ruelle theory shows that the transformation laws under P, C, or T for the fields determine corresponding transformation laws for the in- and out-fields. For example, if there is a scalar particle in the theory, then the associated asymptotic fields satisfy

$$U(I_s)\varphi^{\text{in}}(x)U(I_s)^{-1} = \varphi^{\text{in}}(I_s x)$$
$$U(C)\varphi^{\text{in}}(x)U(C)^{-1} = \varphi^{\text{in}*}(x)$$
$$U(I_t)\varphi^{\text{in}}(x)U(I_t)^{-1} = \varphi^{\text{out}}(I_t x)$$

and also the same equations with φ^{in} interchanged with φ^{out}. If in the theory there is a particle with higher spin, its covariance properties are determined by those of the monomial in the basic fields in the theory which has a non-zero matrix element between the vacuum and the state consisting of one such particle. The operator $U(I_s)$ in (1–45) was chosen so that it reverses the momenta of the particles in the in-states; the same operator has the same effect on the momenta of the particles in the out-states, as is required physically for the parity operator [that is, it is independent of time; cf. the remark after (1–3)]. In the same way $U(C)$ takes particles to their anti-particles. It follows that if $U(I_s)$, $U(C)$ define symmetries they commute with the S-operator, giving rise to a *conservation law:* if Ψ^{in} is an eigenstate of $U(I_s)$, then so is $\Psi^{\text{out}} = S\Psi^{\text{in}}$, and with the same eigenvalue. Invariance under I_t is more complicated, since $U(I_t)$ takes an in-state to an out-state with reversed momenta and spins; it leads to reality conditions on the S-matrix elements.

We now extend the proof of the reconstruction theorem (Theorem 3–7) to theories with a discrete symmetry which gives rise to equations like (3–38) and (3–39). The construction of the corresponding $U(C)$ or Θ follows the same lines as the construction of $U(a, \Lambda)$ in Theorem 3–7. We content ourselves with a discussion of the PCT operator. If we are given further identities among the $\mathscr{W}$'s, for P, C, and T separately, such as (3–38) for C, we can prove the existence of the corresponding operator in the same way.

Theorem 3–9

Consider a field theory of fields $\varphi_{(\alpha)(\dot\beta)}, \ldots \psi_{(\mu)(\dot\nu)}$ where $(\alpha), \ldots (\mu)$ are collections of undotted indices and $(\dot\beta), \ldots (\dot\nu)$ are collections of dotted indices. Suppose (3–39) holds for all vacuum expectation values of the fields, that is,

$$(\Psi_0, \varphi_{(\alpha)(\dot\beta)}(x_1) \ldots \psi_{(\mu)(\dot\nu)}(x_n)\Psi_0)$$
$$= i^F(-1)^J\overline{(\Psi_0, \varphi^*_{(\alpha)(\dot\beta)}(-x_1) \ldots \psi^*_{(\mu)(\dot\nu)}(-x_n)\Psi_0)} \quad (3\text{–}66)$$

where F is the number of half-odd integer spin fields and J is the total number of undotted indices in $(\alpha), \ldots (\nu)$. Then there exists a unique (up to a factor) anti-unitary operator Θ in $\mathscr{H}$ such that for any field in the theory

$$\Theta^{-1}\varphi_{(\alpha)(\dot\beta)}(f)\Theta = (-1)^j i^{F^{(\varphi)}}\varphi^*_{(\alpha)(\dot\beta)}(\mathring{f}), \quad (3\text{–}67)$$

where

$$\mathring{f}(x) = \bar{f}(-x), \quad (\alpha) = (\alpha_1, \ldots \alpha_j), \quad (\dot\beta) = (\dot\beta_1, \ldots \dot\beta_k),$$

and

$$F^{(\varphi)} = 0 \quad \text{if } j + k \text{ is even}$$
$$= 1 \quad \text{if } j + k \text{ is odd.}$$

Proof:

We can define

$$\Theta\varphi_{(\alpha)(\dot\beta)}(f_1) \cdots \psi_{(\mu)(\dot\nu)}(f_n)\Psi_0 = \Theta\varphi_{(\alpha)(\dot\beta)}(f_1)\Theta^{-1}\Theta \ldots \Theta\psi_{(\mu)(\dot\nu)}(f_n)\Theta^{-1}\Theta\Psi_0$$
$$= (-i)^F(-1)^J\varphi^*_{(\alpha)(\dot\beta)}(\mathring{f}_1) \cdots \psi^*_{(\mu)(\dot\nu)}(\mathring{f}_n)\Psi_0 \in H.$$

This equation can be extended by anti-linearity to the whole of the space H (defined in Section 3–4). So defined, Θ is anti-unitary, by virtue of (3–66), which leads to

$$(\Theta f, \Theta g) = \overline{(f, g)}$$

for all finite sequences $f = (f_0, f_1, \ldots)$, $g = (g_0, g_1, \ldots)$ of test functions. In fact Θ defines a mapping of the quotient space H/H_0 obtained by identifying vectors whose difference has zero norm in H (see Section 3–4). For suppose $f \sim g$ in the sense that $\|f - g\| = 0$. This means that a certain sum of vacuum expectation values smeared with the

f's and g's vanishes. We may replace these vacuum expectation values by the PCT transformed ones, using (3–66). The new equation is then a statement that $\|\Theta f - \Theta g\| = 0$, i.e., $\Theta f \sim \Theta g$. Therefore Θ defines an anti-unitary operator on H/H_0, which may be regarded as a subset of $\mathscr{H}$. The extension to all of $\mathscr{H}$ is done by continuity, and, so defined, is anti-unitary. ▮

Theorem 3–9 shows that the presence of PCT symmetry in a field theory is equivalent to the validity of the identities (3–66). It is shown in the next chapter that the identities (3–66) hold in every local field theory. This is obviously a very important result.

BIBLIOGRAPHY

The axioms given in Section 3–1 are due to Gårding and Wightman in

1. A. S. Wightman, *Les Problèmes mathématiques de la théorie quantique des champs*, pp. 11–19, Centre National de la Recherche Scientifique, Paris, 1959; and also
2. A. S. Wightman and L. Gårding, "Fields as Operator-Valued Distributions in Quantum Field Theory," *Ark. Fys.*, **28**, 129 (1964).

The independence of the various axioms in quantum field theory was studied by

3. R. Haag and B. Schroer, "The Postulates of Quantum Field Theory," *J. Math. Phys.*, **3**, 248 (1962).

The generalized free field was introduced by O. W. Greenberg; see for example

4. O. W. Greenberg, "Generalized Free Fields and Models of Local Field Theory," *Ann. Phys.*, **16**, 158 (1961).

The fundamental work of D. Ruelle on collision theory referred to in Section 3–1 is

5. D. Ruelle, "On the Asymptotic Condition in Quantum Field Theory," *Helv. Phys. Acta*, **35**, 34 (1962). Here one finds references to earlier work of Haag and others which laid the groundwork for this form of collision theory. This paper also introduced the definition (3–8) of irreducibility as used in this book.

Our treatment of vacuum expectation values follows

6. A. S. Wightman, "Quantum Field Theory in Terms of Vacuum Expectation Values," *Phys. Rev.*, **101**, 860 (1956).

The important connection of the cluster decomposition property with the uniqueness of the vacuum state was found by

7. K. Hepp, R. Jost, D. Ruelle, and O. Steinmann, "Necessary Condition on Wightman Functions," *Helv. Phys. Acta*, **34**, 542 (1961); see also

8. H. J. Borchers, "On the Structure of the Algebra of Field Observables," *Nuovo Cimento*, **24**, 214 (1962).

The cluster decomposition theorem itself comes in many variants. The paper of D. Ruelle, Ref. 5, contains one, probably the most refined. Others, which also contain references to earlier papers, are

9. H. Araki, "On the Asymptotic Behavior of Vacuum Expectation Values at Large Spacelike Separations," *Ann. Phys.*, **11**, 260 (1960).
10. R. Jost and K. Hepp, "Über die Matrixelemente des Translations Operators," *Helv. Phys. Acta*, **35**, 34 (1962).
11. H. Araki, K. Hepp, and D. Ruelle, "On the Asymptotic Behavior of Wightman Functions in Space-like Directions, *Helv. Phys. Acta*, **35**, 164 (1962).

The last article contains a detailed discussion of the case where there is no mass gap.

For a review of work on the holomorphy domains of vacuum expectation values, see

12. A. O. G. Källén, "Properties of Vacuum Expectation Values of Field Operators," pp. 389–447 in *Dispersion Relations and Elementary Particles*, Wiley, New York, 1960; or
13. A. S. Wightman, "Quantum Field Theory and Analytic Functions of Several Complex Variables," *Proc. Indian Math. Soc.*, **24**, 625 (1960).

The proof of the reconstruction theorem of Section 3–4 is not as close to the original (Ref. 6, above) as to the version of

14. W. Schmidt and K. Baumann, "Quantentheorie der Felder als Distributionstheorie," *Nuovo Cimento*, **4**, 860 (1956).

A general survey of recent developments, including the Haag–Ruelle theory, is to be found in

15. A. S. Wightman, "Recent Achievements of Axiomatic Field Theory," *Proceedings of the Summer Seminar of IAEA*, Trieste, 1962, published as a book, *Theoretical Physics*, IAEA, Vienna, 1963.

The most systematic approach to the non-linear program is due to K. Symanzik, summarized in

16. K. Symanzik, "Green's Functions and the Quantum Theory of Fields," *Lectures in Theoretical Physics III*, Boulder, 1960, pp. 490–531, Interscience, New York, 1961.

For a detailed study of the significance of the phases in the transformation laws of fields under inversions, see

17. G. Feinberg and S. Weinberg, "On the Phase Factors in Inversions," *Nuovo Cimento*, **14**, 571 (1959).

CHAPTER 4

SOME GENERAL THEOREMS OF RELATIVISTIC QUANTUM FIELD THEORY

He had bought a large map representing the sea,
Without the least vestige of land:
And the crew were much pleased when they found it to be
A map they could all understand.

Fit the Second, *The Hunting of the Snark*
LEWIS CARROLL

In the preceding chapters, we have defined what is meant by a relativistic quantum field theory and assembled some tools to aid in the analysis of its structure. In the present chapter, these are used to establish a series of general properties of relativistic quantum field theories.

4–1. THE GLOBAL NATURE OF LOCAL COMMUTATIVITY

Local commutativity asserts the vanishing of commutators or anticommutators, $[\varphi(x), \psi(y)]_{\pm}$, for all space-like $x - y$. An assumption which is apparently weaker is that this condition holds in some smaller region, say $(x - y)^2 < -a < 0$. Theorem 4–1 asserts that such apparently weaker assumptions are in fact not weaker, since local commutativity can be deduced from them.

We prove the theorem for a single hermitian scalar field and the commutator. The proof for an arbitrary set of fields with anticommutators or commutators is just a matter of adjoining a few suffices and changing a few signs, a task we leave to the reader.

Theorem 4–1

Let φ be a hermitian scalar field satisfying axioms I and II, but, instead of satisfying axiom III, having the property

$$[\varphi(x), \varphi(y)]_{-} = 0 \qquad (4\text{–}1)$$

for x and y varying in some space-like separated open sets. Suppose the vacuum is cyclic with respect to φ. Then φ is local, that is, (4–1) holds for all space-like separated x and y.

Remarks:

1. Because of the transformation law of the field we have

$$U(a, \Lambda)[\varphi(x), \varphi(y)]_{\pm} U(a, \Lambda)^{-1} = [\varphi(\Lambda x + a), \varphi(\Lambda y + a)]_{\pm}$$

so the vanishing of (4–1) in the neighborhood of a point $\{x, y\}$ implies its vanishing in a neighborhood of $\{\Lambda x + a, \Lambda y + a\}$. Thus, the assumption of the theorem immediately implies the vanishing of the commutator for all $\{x, y\}$ such that $(x - y)^2$ is in some open set of the negative real axis.

2. The asserted vanishing of the commutator means that this operator vanishes on the domain of definition D of φ.

Proof:

Consider the two distributions F_1 and F_2, where F_1 is defined by

$$\begin{aligned} F_1(x_1 - x_2, x_2 - x_3, \ldots x_{j-1} - x_j, x_j - x, x - y, y - y_1, \\ y_1 - y_2, \ldots y_{k-1} - y_k) \\ = (\Psi_0, \varphi(x_1) \ldots \varphi(x_j)\varphi(x)\varphi(y)\varphi(y_1) \ldots \varphi(y_k)\Psi_0) \quad (4\text{–}2) \end{aligned}$$

and F_2 is obtained from it by interchange of x and y. They are respectively boundary values of holomorphic functions F_1 and F_2. F_1 is holomorphic in the extended tube $\mathscr{T}'_{j+k+1}$ in the variables

$$\begin{aligned} x_1 - x_2 - i\eta_1, \ldots x_{j-1} - x_j - i\eta_{j-1}, x_j - x - i\eta, x - y - i\eta', \\ y - y_1 - i\eta'', y_1 - y_2 - i\rho_1, \ldots y_{k-1} - y_k - i\rho_{k-1}. \quad (4\text{–}3) \end{aligned}$$

To get F_1 as a boundary value one has to let $\eta_1, \ldots \eta_{j-1}, \eta, \eta', \eta'', \rho_1, \ldots \rho_{k-1} \to 0$ in V_+. The same holds for F_2 with x and y interchanged in the arguments. However, if we want to regard F_2 as a function of the arguments (4–3), then those arguments have to vary over the permuted extended tube $\mathscr{P}\mathscr{T}'_{j+k+1}$, where $\mathscr{P}$ is the transformation which carries (4–3) into

$$\begin{aligned} x_1 - x_2 - i\eta_1, \ldots x_{j-1} - x_j - i\eta_{j-1}, (x_j - x - i\eta) + (x - y - i\eta'), \\ -(x - y - i\eta'), (x - y - i\eta') + (y - y_1 - i\eta''), \\ y_1 - y_2 - i\rho_1, \ldots y_{k-1} - y_k - i\rho_{k-1}. \quad (4\text{–}4) \end{aligned}$$

It is this second way of looking at F_2 which will be important for the following.

Now by the discussion at the end of Section 2–4, we know that $\mathscr{T}'_{j+k+1}$ and $\mathscr{P}\mathscr{T}'_{j+k+1}$ possess real points in common and, in fact, open sets of real points for which $x - y$ lies in a given neighborhood of space-like vectors. In such a neighborhood $F_1 = F_1$ and $F_2 = F_2$. Thus, the hypothesis of the present theorem implies $F_1 = F_2$ in a real open set, $\mathscr{U}$, and therefore in a complex open set.

If we were to proceed hastily at this point, we would argue as follows. The equality of F_1 and F_2 holds throughout $\mathscr{T}'_{j+k+1} \cup \mathscr{P}\mathscr{T}'_{j+k+1}$ by analytic continuation. Thus we can pass to boundary values and get $\boldsymbol{F}_1 = \boldsymbol{F}_2$ for all space-like $x - y$. This argument is invalid for two reasons. First, although we know F_1 is single-valued throughout $\mathscr{T}'_{j+k+1}$ and F_2 throughout $\mathscr{P}\mathscr{T}'_{j+k+1}$, we have not shown that they are single-valued throughout $\mathscr{T}'_{j+k+1} \cup \mathscr{P}\mathscr{T}'_{j+k+1}$. Second, if in the relation

$$F_1(\zeta_1,\ldots\zeta_{j+k+1}) = F_2(\zeta_1,\ldots\zeta_{j-2}, \zeta_{j-1} + \zeta_j, -\zeta_j, \zeta_j + \zeta_{j+1},\ldots\zeta_{j+k+1}) \tag{4-5}$$

we approach the boundary in $\mathscr{T}_{j+k+1}$ so that the imaginary parts of $\zeta_1\ldots\zeta_{j+k+1}$ lie within the appropriate cone making $F_1 \to \boldsymbol{F}_1$, then on the right-hand side we shall not be approaching the boundary in $\mathscr{T}_{j+k+1}$, so that we cannot be sure that $F_2 \to \boldsymbol{F}_2$. We shall avoid these objections by a more careful argument. A different method from the one given here is to be found in Ref. 1; it does not use the edge of the wedge theorem.

Continuing with the proof, by the first remark we may assume that $\mathscr{U}$ depends on $x - y$ only through $(x - y)^2$. We can continue the relation $F_1 = F_2$ along the path given by points of the form

$$\rho(x_1 - x_2),\ldots\rho(x_{j-1} - x_j),\ \rho(x_j - x),\ \rho(x - y),\ \rho(y - y_1),$$
$$\rho(y_1 - y_2),\ldots\rho(y_{k-1} - y_k),\ \rho > 0, \tag{4-6}$$

where the point with $\rho = 1$ lies in the neighborhood $\mathscr{U}$ introduced above; it can easily be seen that this curve lies in the common set of Jost points of the extended tubes $\mathscr{T}'_{j+k+1}$ and $\mathscr{P}\mathscr{T}'_{j+k+1}$ if the point with $\rho = 1$ does. Clearly, we can take the vector $x - y$ to a point with any negative value of $\xi^2 = \rho^2(x - y)^2$ whatever, the open set $\mathscr{U}$ being transformed to another open set $\rho\mathscr{U}$. Therefore if $(x - y)^2 < 0$,

$$\boldsymbol{F}_1 = \boldsymbol{F}_2$$

in some open set $\rho\mathscr{U}$. (Clearly, this argument avoids the first pitfall mentioned in the preceding paragraph, the possible several-valuedness of F_1 and F_2 in $\mathscr{T}'_{j+k+1} \cup \mathscr{P}\mathscr{T}'_{j+k+1}$.)

The last step in the proof shows that the equality $\boldsymbol{F}_1 = \boldsymbol{F}_2$, if $(x - y)^2 < 0$, and all the variables lie in $\rho\mathscr{U}$, implies its validity for all $x_1,\ldots x_j$, $y_1,\ldots y_k$ and all space-like $x - y$. For this purpose we introduce two new tempered distributions, f_1 and f_2. Let $(x' - y')$ such that $(x' - y')^2 < 0$ be given and let $\rho\mathscr{U}$ be the corresponding open set. The equation

$$f_1(x_1 - x_2,\ldots x_{j-1} - x_j, x_j, -y_1, y_1 - y_2,\ldots y_{k-1} - y_k)$$
$$= \int h(x, y)\, dx\, dy(\Psi_0, \varphi(x_1)\ldots\varphi(x_j)\varphi(x)\varphi(y)\varphi(y_1)\ldots\varphi(y_k)\Psi_0) \tag{4-7}$$

defines f_1, and f_2 is defined in the same way except that $\varphi(x)$ and $\varphi(y)$ are interchanged. Here the test function h has a compact support K containing the given point $x' - y'$, the support K being small enough so that, for every vector $x - y$ in K over which we are integrating in (4–7), the set $\rho\mathscr{U}$ always contains a fixed open set V in the space of the remaining variables $x_1 - x_2, \ldots x_{j-1} - x_j, x_j, -y_1, y_1 - y_2, \ldots y_{k-1} - y_k$.

The distributions f_1 and f_2 are, respectively, boundary values of holomorphic functions $\boldsymbol{f}_1$ and $\boldsymbol{f}_2$ holomorphic when their arguments vary over the tube $\mathscr{T}_{j+k+1}$ in the variables $x_1 - x_2 - i\eta_1, \ldots x_{j-1} - x_j - i\eta_{j-1}, x_j - i\eta, -y_1 - i\eta', y_1 - y_2 - i\rho_1, \ldots y_{k-1} - y_k - i\rho_{k-1}$. This follows from Theorem 2–6, because the Fourier transforms of f_1 and f_2 vanish unless their arguments lie in the physical spectrum. Now we apply Theorem 2–17 to the difference $\boldsymbol{f}_1 - \boldsymbol{f}_2$. It is holomorphic in $\mathscr{T}_{j+k+1}$ and vanishes for an open set of real points; therefore it vanishes identically and so does the boundary value $f_1 - f_2$. Since the test function h is arbitrary provided its support K lies in a sufficiently small neighborhood of $x' - y'$, we conclude $F_1 = F_2$ in some neighborhood of $x' - y'$. Since $x' - y'$ was an arbitrary space-like vector we conclude† $F_1 = F_2$ if $(x - y)^2 < 0$. This means

$$(\Phi, [\varphi(x), \varphi(y)]\Psi) = 0 \qquad \text{for } (x - y)^2 < 0,\ \Psi, \Phi \in D_0, \tag{4–8}$$

the domain of vectors obtained by applying the smeared fields to the vacuum. Since D_0 is dense (the cyclicity axiom), and the scalar product is continuous, we get (4–8) for all $\Phi \in D$, $\Psi \in D_0$. Using the hermiticity relation

$$-(\Phi, [\varphi(x), \varphi(y)]\Psi) = ([\varphi(x), \varphi(y)]\Phi, \Psi),$$

we see that if $\Phi \in D$, $(x - y)^2 < 0$

$$[\varphi(x), \varphi(y)]\Phi \quad \text{is orthogonal to all} \quad \Psi \in D_0$$

and therefore vanishes. ▮

4–2. PROPERTIES OF THE POLYNOMIAL ALGEBRA OF AN OPEN SET

Quantum field theory provides a set of candidates for local measurements, observables which correspond to field measurements performed in a laboratory of finite size and completed in a finite time. These are the field operators smeared with test functions of compact support. The present section is concerned with the properties of an algebra associated with such quantities. For simplicity, we again discuss the theory of a hermitian scalar field.

† See Section 2–1.

Let $\mathscr{P}(\mathcal{O})$ be the set of all polynomials of the form

$$c + \sum_{j=1}^{N} \varphi(f_1^{(j)}) \dots \varphi(f_j^{(j)}) \tag{4-9}$$

where the $f_k^{(j)}$ are test functions whose support is in the open set $\mathcal{O}$ of space-time, and c is any complex constant. Clearly, if p and q are two such polynomials so are $p + q$, αp, p^*, and pq. This means that $\mathscr{P}(\mathcal{O})$ is a $*$ algebra, the *polynomial algebra of* $\mathcal{O}$. The next theorem is a surprise: the algebra associated with *any* open set $\mathcal{O}$ has Ψ_0 as a cyclic vector. It is due to Reeh and Schlieder.

Theorem 4–2

Suppose $\mathcal{O}$ is an open set of space-time. Then Ψ_0 is a cyclic vector for $\mathscr{P}(\mathcal{O})$, if it is a cyclic vector for $\mathscr{P}(\mathbf{R}^4)$. That is, vectors of the form

$$\sum_{j=0}^{N} \varphi(f_1^{(j)}) \dots \varphi(f_j^{(j)}) \Psi_0 \tag{4-10}$$

with $\operatorname{supp} f_j^{(k)} \subset \mathcal{O}$ are dense in $\mathscr{H}$.

Proof:

Let Ψ be orthogonal to all vectors of the form (4–10). We are going to prove that Ψ is also orthogonal to all polynomials in the smeared field with no conditions on the supports of the test functions, that is, that Ψ is orthogonal to $\mathscr{P}(\mathbf{R}^4)\Psi_0 = D_0$. This will then imply that $\Psi = 0$, because by assumption D_0 spans the Hilbert space $\mathscr{H}$. The method of proof is another typical application of the principle of analytic continuation.

The first step in the proof is to argue that the symbolic expression

$$(\Psi, \Psi_f) = \int (\Psi, \varphi(x_1)\varphi(x_2) \dots \varphi(x_n)\Psi_0) f(x_1, x_2, \dots x_n)\, dx_1 \dots dx_n$$

has a meaning for all test functions, f, in $\mathscr{S}$, and any Ψ in $\mathscr{H}$. The proof goes just like that in the remarks after (3–24). One first notes that $(\Psi, \varphi(f_1) \dots \varphi(f_n)\Psi_0)$ is a continuous multilinear functional of the test functions $f_1 \dots f_n$ and applies the Schwartz nuclear theorem to extend it to a tempered distribution in all the variables together. Thus there is a tempered distribution F defined by

$$F(-x_1, x_1 - x_2, \dots x_{n-1} - x_n) \equiv (\Psi, \varphi(x_1) \dots \varphi(x_n)\Psi_0).$$

Standard arguments† show that the Fourier transform of F vanishes unless each four-momentum variable lies in the physical spectrum.

† See Section 2–6.

Therefore there exists a holomorphic function F, holomorphic in the tube $\mathscr{T}_n$ in the variables $(-x_1) - i\eta_1, (x_1 - x_2) - i\eta_2, \ldots (x_{n-1} - x_n) - i\eta_n$, whose boundary value as $\eta_1 \ldots \eta_n \to 0$ in V_+ is F. This boundary value vanishes for $-x_1, x_1 - x_2, \ldots x_{n-1} - x_n$ in an open set determined by $x_1 \ldots x_n \in \mathscr{O}$. Thus, by Theorem 2–17 F vanishes and therefore its boundary value vanishes for all $x_1, \ldots x_n$. This shows Ψ is orthogonal to D_0. ▌

It is well known in the theory of algebras of bounded operators that a cyclic vector for a $*$-algebra, $\mathscr{P}$, is a separating vector for its commutant. That means: if the set of vectors $\mathscr{P}\Psi$ is dense, then the equality $T\Psi = 0$ for an operator T commuting with all operators of $\mathscr{P}$ implies $T = 0$. (For a proof see Ref. 4.) Theorem 4–3 is an analogue of this statement for our case. It can be interpreted as meaning that it is difficult to isolate a system described by fields from outside effects. The algebra $\mathscr{P}(\mathscr{O})$ never contains any annihilation or creation operators if the open set satisfies a certain boundedness condition. To define the boundedness condition, consider the set of all points which are space-like with respect to every point of $\mathscr{O}$. The interior of this set is $\mathscr{O}'$. Then we assert

Theorem 4–3

If $\mathscr{O}$ is an open set for which $\mathscr{O}'$ is not empty, and $T \in \mathscr{P}(\mathscr{O})$, then

$$T\Psi_0 = 0 \tag{4–11}$$

implies $T = 0$.

Proof:

Let Φ be a vector in the domain of definition D of φ. Let $\Psi = P'\Psi_0$, where $P' \in \mathscr{P}(\mathscr{O}')$. Then for any $T \in \mathscr{P}(\mathscr{O})$ satisfying (4–11),

$$(\Psi, T^*\Phi) = (T\Psi, \Phi) = (TP'\Psi_0, \Phi) = (P'T\Psi_0, \Phi) = 0 \tag{4–12}$$

where in the last equality (4–11) has been used. Since, by Theorem 4–2, vectors of the form Ψ span $\mathscr{H}$, we have $T^*\Phi = 0$. This in turn implies $T\Psi = 0$, for $\Psi \in D$ since $(T\Psi, \Phi) = (\Psi, T^*\Phi)$ and D is dense. ▌

Remarks:

1. Theorem 4–3 retains its validity if, instead of assuming that commutators of fields vanish at space-like separations, we assume that anti-commutators vanish. This will be used in the proof of Theorem 4–8.

2. Any bounded open set $\mathcal{O}$ has the property that $\mathcal{O}'$ is non-empty, and so the theorem applies.

Theorem 4–4, also due to Reeh and Schlieder, shows that $\mathscr{P}(\mathcal{O})$ becomes irreducible when a single operator is adjoined.

Theorem 4–4

Let E_0 be the projection operator onto the vacuum state, which is supposed cyclic with respect to the field. Then, for every open set $\mathcal{O}$, the set of operators $\{E_0, \mathscr{P}(\mathcal{O})\}$ is irreducible.

Proof:

Recall the definition of an irreducible set, Eq. (3–8). Suppose that for some bounded operator, C, and all $\Phi, \Psi \in D_0$, we have

$$(\Phi, C\varphi(f)\Psi) = (\varphi(f)^*\Phi, C\Psi), \tag{4–13}$$

where $\operatorname{supp} f \subset \mathcal{O}$, and suppose that

$$CE_0 = E_0 C. \tag{4–14}$$

Then (4–13) holds, in particular, for a state Ψ of the form

$$\Psi = p\Psi_0, \qquad p \in \mathscr{P}(\mathcal{O}) \tag{4–15}$$

and

$$\begin{aligned}(\Phi, C\Psi) &= (\Phi, Cp\Psi_0) = (p^*\Phi, C\Psi_0) = (p^*\Phi, CE_0\Psi_0) \\ &= (p^*\Phi, E_0C\Psi_0) = (p^*\Phi, \Psi_0)(\Psi_0, C\Psi_0)\end{aligned} \tag{4–16}$$

where (4–13), (4–14), and the definition of E_0,

$$E_0\Psi = (\Psi_0, \Psi)\Psi_0,$$

have been used. Clearly, (4–16) says

$$(\Phi, C\Psi) = c_0(\Phi, \Psi),$$

where

$$c_0 = (\Psi_0, C\Psi_0).$$

Since a bounded operator is continuous and states Φ, Ψ are dense in $\mathscr{H}$, this implies $C = c_0$. ∎

Theorem 4–4 shows that for any $\mathcal{O}$ the adjunction of E_0 to $\mathscr{P}(\mathcal{O})$ produces an irreducible set of operators. The next theorem shows that for a special choice of $\mathcal{O}$, namely, the whole of space-time, E_0 is effectively contained in $\mathscr{P}(\mathcal{O})$, so $\mathscr{P}(\mathcal{O})$ itself is irreducible. We give a proof for a theory of a neutral scalar field. The extension to a set of fields of arbitrary transformation laws is easy.

Theorem 4–5

In any field theory, the smeared fields form an irreducible set of operators.

Remark:

Recall that with our definition of a field theory the vacuum is a cyclic vector for the smeared fields; this hypothesis is essential for the truth of the theorem. The theorem provides the promised justification of the definition of field theory in terms of cyclicity of the vacuum rather than irreducibility of the smeared fields.

Proof:

Notice that if C is a bounded operator which satisfies

$$(\Phi, C\varphi(f)\Psi) = (\varphi(f)^*\Phi, C\Psi)$$

for all $\Phi, \Psi \in D_0$, then C satisfies

$$(\Phi, C\varphi(f_1)\ldots\varphi(f_n)\Psi) = (\varphi(f_n)^*\ldots\varphi(f_1)^*\Phi, C\Psi). \qquad (4\text{–}17)$$

Consider, in particular, the case

$$(\Psi_0, C\varphi(\{a, 1\}f_1)\ldots\varphi(\{a, 1\}f_n)\Psi_0)$$
$$= (\varphi(\{a, 1\}f_n)^*\ldots\varphi(\{a, 1\}f_1)^*\Psi_0, C\Psi_0). \qquad (4\text{–}18)$$

Rewritten using the transformation law of φ and the invariance of the vacuum state under translation, it is

$$(\Psi_0, CU(a, 1)\varphi(f_1)\ldots\varphi(f_n)\Psi_0) = (\varphi(f_n)^*\ldots\varphi(f_1)^*\Psi_0, U(-a, 1)C\Psi_0).$$

By Fourier-transforming this equation in a we get the assertion

$$(\Psi_0, CE(S)\varphi(f_1)\ldots\varphi(f_n)\Psi_0) = (\varphi(f_n)^*\ldots\varphi(f_1)^*\Psi_0, E(-S)C\Psi_0),$$

where S is any measurable set in momentum space. This is explained in Section 2–6.

But since the spectrum of energy momentum lies in or on the plus cone, if $-S$ lies in the physical spectrum and does not include $p = 0$, the left-hand side vanishes. That means that $C\Psi_0$ is orthogonal to all states $E(-S)\varphi(f_n)^*\ldots\varphi(f_1)^*\Psi_0$ and therefore $C\Psi_0 = c\Psi_0$, where c is a complex number. Then (4–17) for the special case $\Psi = \Psi_0$ is

$$(\Phi, C\varphi(f_1)\ldots\varphi(f_n)\Psi_0) = (\varphi(f_n)^*\ldots\varphi(f_1)^*\Phi, C\Psi_0)$$
$$= c(\Phi, \varphi(f_1)\ldots\varphi(f_n)\Psi_0).$$

This implies

$$(\Phi, C\Psi) = c(\Phi, \Psi)$$

for all $\Phi, \Psi \in D_0$, whence by the continuity of C and the denseness of D_0, $C = c$. ▮

It should be pointed out that although the vacuum is cyclic for all algebras $\mathscr{P}(\mathcal{O})$ they are not all irreducible. For example, if $\mathcal{O} \neq \mathbf{R}^4$, we cannot prove (4–18) for all a if we are given (4–17) only for test functions with support in $\mathcal{O}$. In particular, we cannot prove the time slice axiom (Section 3–2). For example, in a theory of nucleons interacting with π-mesons (we assume the particles are stable) we may introduce fields ψ_p, ψ_n and $\varphi^\pm$, φ^0 for the p, n and $\pi^\pm$, π^0, respectively. In an asymptotically complete theory, states of the form $P(\psi_p, \psi_n)\Psi_0$ span the Hilbert space (including π-mesons). This follows from the Haag–Ruelle theory if $P(\psi_p, \psi_n)\Psi_0$ is not orthogonal to the one-meson states. Theorem 4–5 then asserts that the operators (ψ_p, ψ_n) form an irreducible set, when smeared with arbitrary test functions. The same remarks show that (ψ_p, φ^-) and (ψ_n, φ^+) are always alternative irreducible sets. However as far as we know it could happen that in one theory (ψ_p, ψ_n) is irreducible in a time slice, while in another theory it is reducible, while $(\psi_p, \psi_n, \varphi)$ is an irreducible set. This subject is closely related to the question of whether the pion is in some sense a bound state of a nucleon and anti-nucleon, or not.

A natural continuation of the arguments of this section would lead to the introduction of the notion of the von Neumann algebra $\mathscr{R}(\mathcal{O})$ of an open set $\mathcal{O}$. This is a $*$-algebra of *bounded* operators. The most natural way to obtain them would be to use the spectral resolution of the hermitian elements of $\mathscr{P}(\mathcal{O})$, and to take the algebra generated by their spectral projections. We shall not undertake to explain these developments here, but only remark that there are good reasons to believe that the study of the $\mathscr{R}(\mathcal{O})$ is worthwhile. There are indications that two field theories with the same representation of the Lorentz group will yield the same S matrix if and only if their $\mathscr{R}(\mathcal{O})$ are isomorphic.† This lends interest to the theorems of the present section.

4–3. THE PCT THEOREM

In Section 3–5 we saw that the existence of a PCT operator, Θ, for a set of fields was equivalent to the validity of the identities (3–66):

$$(\Psi_0, \varphi_1(x_1)\ldots\varphi_n(x_n)\Psi_0) = (-1)^J i^F(\Psi_0, \varphi_n(-x_n)\ldots\varphi_1(-x_1)\Psi_0). \tag{4–19}$$

In this section we shall prove that the identities (4–19) hold in every field theory of local fields. That is the PCT theorem or Lüders–Pauli theorem. In fact, a more precise result will be proved which shows that a weaker condition, so-called *weak local commutativity*, is sufficient

† See Ref. 3 of Chapter 3.

for the validity of the identities (4–19). This refined form which we also call the *PCT* theorem is due to Jost, as is the method of proof.

For the sake of clarity, we shall first state and prove the *PCT* theorem for the theory of a neutral scalar field and only afterwards extend it to the case of general fields.

Theorem 4–6 (*PCT Theorem for a Neutral Scalar Field*)

Let φ be a hermitian scalar field satisfying axioms I and II but not necessarily III (*LC*). If the *PCT* condition

$$(\Psi_0, \varphi(x_1)\ldots\varphi(x_n)\Psi_0) = (\Psi_0, \varphi(-x_n)\ldots\varphi(-x_1)\Psi_0) \quad (4\text{–}20)$$

holds for all $x_1\ldots x_n$, then for every $x_1\ldots x_n$ such that $x_1 - x_2, \ldots x_{n-1} - x_n$ is a Jost point, the W(eak) L(ocal) C(ommutativity) condition

$$(\Psi_0, \varphi(x_1)\ldots\varphi(x_n)\Psi_0) = (\Psi_0, \varphi(x_n)\ldots\varphi(x_1)\Psi_0) \quad (4\text{–}21)$$

holds.

Conversely, if the *WLC* condition (4–21) holds in a (real) neighborhood of a Jost point, then the *PCT* condition (4–20) holds everywhere.

Since *LC* implies the *WLC* conditions, every field theory of a local hermitian scalar field has *PCT* symmetry.

Proof:

Our first step is to translate the *PCT* condition (4–20) into an equivalent relation for a holomorphic function. By Theorem 3–5, we know that there is a holomorphic function W of the $n-1$ complex vectors $\zeta_1, \ldots \zeta_{n-1}$, where $\zeta_j = \xi_j - i\eta_j$ $\xi_j = x_j - x_{j+1}$, holomorphic in the extended tube $\mathscr{T}'_{n-1}$ and such that

$$\lim_{\substack{\eta_1,\ldots\eta_{n-1}\to 0\\ \in V_+}} W(\zeta_1,\ldots\zeta_{n-1}) = W(\xi_1,\ldots\xi_{n-1}) = (\Psi_0, \varphi(x_1)\ldots\varphi(x_n)\Psi_0). \quad (4\text{–}22)$$

Furthermore, we know by Theorem 3–5 that W is invariant under proper complex Lorentz transformations

$$W(\zeta_1,\ldots\zeta_{n-1}) = W(\Lambda\zeta_1,\ldots\Lambda\zeta_{n-1}), \quad (4\text{–}23)$$

for $\Lambda \in L_+(C)$, and all $\zeta_1,\ldots\zeta_{n-1} \in \mathscr{T}'_{n-1}$. This implies, in particular,

$$W(\zeta_1,\ldots\zeta_{n-1}) = W(-\zeta_1,\ldots -\zeta_{n-1}) \text{ for } \zeta_1,\ldots\zeta_{n-1} \in \mathscr{T}'_{n-1}, \quad (4\text{–}24)$$

a relation that we shall use shortly. The right-hand side of (4–20) is also a boundary value of a holomorphic function:

$$\lim_{\substack{\eta_1,\dots\eta_{n-1}\to 0\\ \in V_+}} W(\zeta_{n-1},\dots\zeta_1) = W(\xi_{n-1},\dots\xi_1)$$

$$= (\Psi_0, \varphi(-x_n)\varphi(-x_{n-1})\dots\varphi(-x_1)\Psi_0). \quad (4\text{–}25)$$

Since $W(\zeta_1,\dots\zeta_{n-1}) - W(\zeta_{n-1},\dots\zeta_1)$ is holomorphic throughout $\mathscr{T}_{n-1}$ and vanishes for $\zeta_1,\dots\zeta_{n-1}$ real by (4–20) we see by Theorem 2–17 that (4–20) implies

$$W(\zeta_1,\dots\zeta_{n-1}) = W(\zeta_{n-1},\dots\zeta_1). \quad (4\text{–}26)$$

But, conversely, if (4–26) holds in an arbitrary neighborhood of a point of $\mathscr{T}'_{n-1}$, it holds everywhere in $\mathscr{T}'_{n-1}$ and, by passing to the boundary in the tube $\mathscr{T}_{n-1}$, one recovers (4–20). Thus, (4–20) is completely equivalent to the relation (4–26) for the holomorphic function W.

Now we combine the relations (4–24) and (4–26) to get

$$W(\zeta_1,\dots\zeta_{n-1}) = W(-\zeta_{n-1},\dots -\zeta_1) \quad (4\text{–}27)$$

valid throughout $\mathscr{T}'_{n-1}$. If we attempted to pass to the boundary in this relation, we would not get a relation between vacuum expectation values because if $\zeta_1\dots\zeta_{n-1}$ approached real vectors in the plus tube, $\mathscr{T}_{n-1}$, $-\zeta_{n-1},\dots -\zeta_1$ would approach real vectors in the minus tube. However, at a real point of holomorphy (Jost point), (4–27) *is* a relation between vacuum expectation values, namely,

$$W(\xi_1,\dots\xi_{n-1}) = (\Psi_0, \varphi(x_1)\dots\varphi(x_n)\Psi_0)$$

$$= W(-\xi_{n-1},\dots -\xi_1) = (\Psi_0, \varphi(x_n)\dots\varphi(x_1)\Psi_0), \quad (4\text{–}28)$$

which is just *WLC*. Thus, we have proved the first half of the theorem.

The converse is easy. If (4–28) holds in a real neighborhood of a real point of holomorphy then it holds in a complex neighborhood, and therefore, by analytic continuation (4–27) holds throughout $\mathscr{T}'_{n-1}$. By the complex Lorentz invariance (4–24) this implies the relation (4–26), which we have seen is equivalent to the *PCT* condition on the vacuum expectation value.

The last statement of the theorem is an immediate consequence of the fact that at a Jost point all differences $x_j - x_k$, $j \neq k$ are space-like (see Theorem 2–12), so axiom III (*LC*) clearly implies the *WLC* condition there. ∎

The proof of the *PCT* theorem for a field theory of fields $\varphi_\alpha, \varphi_\beta, \dots$ transforming according to general irreducible representations under the Lorentz group follows a similar pattern. The W are not invariant but instead have a transformation law under $SL(2,C) \otimes SL(2,C)$:

$$\textstyle\sum_{\mu'\ldots\nu'} S^{(\varphi)}_{\mu\mu'}(A, B)\ldots S^{(\psi)}_{\nu\nu'}(A, B)W_{\mu'\ldots\nu'}(\zeta_1,\ldots\zeta_{n-1}) = W_{\mu\ldots\nu}(\Lambda(A, B)\zeta_1,\ldots\Lambda(A, B)\zeta_{n-1}). \quad (4\text{–}29)$$

Here $W_{\mu\ldots\nu}$ is holomorphic in $\mathscr{T}'_{n-1}$ and

$$\lim_{\substack{\eta_1,\ldots\eta_{n-1}\to 0\\ \in V_+}} W_{\mu\ldots\nu}(\zeta_1,\ldots\zeta_{n-1}) = W_{\mu\ldots\nu}(\xi_1,\ldots\xi_{n-1}) = (\Psi_0, \varphi_\mu(x_1)\ldots\psi_\nu(x_n)\Psi_0). \quad (4\text{–}30)$$

Equation (4–29) is the analogue of (4–23). It implies that if the total number of undotted and dotted indices is odd the vacuum expectation value vanishes. To see this, compare (4–29) for $\{A, B\} = \{-1, 1\}$ and $\{1, -1\}$. The right-hand side is the same for these two values, since $\Lambda(-1, 1) = \Lambda(1, -1) = -1$, while the left-hand sides have opposite signs according to (1–27). In any case

$$S^{(\varphi)}_{\mu\mu'}(-1, 1)\ldots S^{(\psi)}_{\nu\nu'}(-1, 1) = \delta_{\mu\mu'}\ldots\delta_{\nu\nu'}(-1)^J, \quad (4\text{–}31)$$

where J is the total number of undotted indices, so the analogue of (4–24) is

$$W_{\mu\ldots\nu}(\zeta_1,\ldots\zeta_{n-1}) = (-1)^J W_{\mu\ldots\nu}(-\zeta_1,\ldots-\zeta_{n-1}) \quad (4\text{–}32)$$

for all $\zeta_1\ldots\zeta_{n-1} \in \mathscr{T}'_{n-1}$.

For these general fields, the analogue of the PCT condition (4–20) is (4–19). This gives rise to the relation between holomorphic functions

$$W_{\mu\ldots\nu}(\zeta_1,\ldots\zeta_{n-1}) = i^F(-1)^J\hat{W}_{\nu\ldots\mu}(\zeta_{n-1},\ldots\zeta_1), \quad (4\text{–}33)$$

where $\hat{W}_{\nu\ldots\mu}$ is holomorphic in $\mathscr{T}'_{n-1}$ and has the boundary value

$$\lim_{\substack{\eta_1,\ldots\eta_{n-1}\to 0\\ \in V_+}} \hat{W}_{\nu\ldots\mu}(\zeta_{n-1},\ldots\zeta_1) = \hat{W}_{\nu\ldots\mu}(\xi_{n-1},\ldots\xi_1) = (\Psi_0, \psi_\nu(-x_n)\ldots\varphi_\mu(-x_1)\Psi_0). \quad (4\text{–}34)$$

The combination of (4–32) and (4–33) gives the analogue of (4–27):

$$W_{\mu\ldots\nu}(\zeta_1,\ldots\zeta_{n-1}) = i^F\hat{W}_{\nu\ldots\mu}(-\zeta_{n-1},\ldots-\zeta_1), \quad (4\text{–}35)$$

which, at a real point in $\mathscr{T}'_{n-1}$, is

$$(\Psi_0, \varphi_\mu(x_1)\ldots\psi_\nu(x_n)\Psi_0) = i^F(\Psi_0, \psi_\nu(x_n)\ldots\varphi_\mu(x_1)\Psi_0). \quad (4\text{–}36)$$

This is the WLC relation for this case and it only remains to verify that it is a consequence of the commutation relations of the fields. We are going to assume "normal" commutation relations. That means that all fields commute at space-like separations except fields whose total number of indices is odd, and they anti-commute. This immediately yields a factor $(-1)^{(F-1)+(F-2)+\cdots 1} = (-1)^{F(F-1)/2} = i^F$, where the last equality holds because F is even in the relevant cases. With the use of the relations (4–29) to (4–36) the proof goes just as before. We summarize:

Theorem 4–7 (*PCT Theorem for General Spin*)

Let $\varphi_\mu \ldots \psi_\nu$ be spinor fields satisfying axioms I and II but not necessarily III (*LC*). If the *PCT* condition

$$(\Psi_0, \varphi_\mu(x_1)\ldots\psi_\nu(x_n)\Psi_0) = i^F(-1)^J(\Psi_0, \psi_\nu(-x_n)\ldots\varphi_\mu(-x_1)\Psi_0)$$

holds for all $x_1 \ldots x_n$, then for every $x_1 \ldots x_n$ such that $x_1 - x_2, \ldots x_{n-1} - x_n$ is a Jost point, the *WLC* condition

$$(\Psi_0, \varphi_\mu(x_1)\ldots\psi_\nu(x_n)\Psi_0) = i^F(\Psi_0, \psi_\nu(x_n)\ldots\varphi_\mu(x_1)\Psi_0)$$

holds.

Conversely, if the *WLC* condition holds in a (real) neighborhood of a Jost point, then the *PCT* conditions holds everywhere.

The normal commutation relations among the $\varphi \ldots \psi$ imply the *WLC* conditions, so every field theory of fields with normal commutation relations has *PCT* symmetry.

We shall see in the next section that if fields have "abnormal" commutation relations there is also a symmetry Θ in the theory but it differs from the definition (1–53) by an extra factor ± 1. This is discussed at the end of the next section.

It should be emphasized that while the assumption of local commutativity implies that a field theory has *PCT* symmetry, only weak local commutativity is necessary for this conclusion. This is important in principle because it is relatively straightforward to construct a field theory which satisfies weak local commutativity (but not local commutativity) and has a non-trivial S-matrix. Such a theory will have *PCT* as a symmetry. The *PCT* symmetry observed in nature therefore is not a strong support for the hypothesis of local commutativity.

4–4. SPIN AND STATISTICS

All experimental evidence indicates that systems with integer spin obey the laws of Bose–Einstein statistics, and systems with half-odd integer spin those of Fermi–Dirac statistics. Although there are perfectly respectable laws of statistics different from either Bose–Einstein or Fermi–Dirac, so far no system has been seen to follow them

(Ref. 28). A natural way to arrive at Bose–Einstein statistics is to describe the system in question by a field which commutes for space-like separations, while the analogous way for Fermi–Dirac statistics is to use a field which anti-commutes for space-like separations. The theorem on the connection of spin with statistics or, as we shall say for brevity, the spin-statistics theorem, is an assertion that in quantum field theory a non-trivial integer spin field cannot have an anti-commutator vanishing for space-like separations, and a non-trivial half-odd integer spin field cannot have a commutator vanishing for space-like separations. If one puts aside the possibility of laws of statistics other than Bose–Einstein or Fermi–Dirac, the spin statistics theorem then accounts for the experimental results.

When one turns from the commutation relations for a given field to those between different fields the situation becomes more complicated. It turns out that "abnormal" commutation relations in which two integer spin fields or an integer spin and a half-odd integer spin field anti-commute, or two half-odd integer spin fields commute, can be realized but, in general, the resulting theories possess special symmetries. By virtue of these symmetries it turns out that there always exists another set of fields, satisfying normal commutation relations and related to the original fields by a so-called Klein transformation. The original theory can equally well be regarded as a theory of this set of fields. In this sense, a theory with abnormal commutation relations is a special case of a theory with normal commutation relations, one which possesses a set of symmetries.

We shall prove these statements in turn. Following Dell'Antonio we first dispose of the possibility that a field component φ has different commutation relations with a field component ψ and its adjoint ψ^*.

Theorem 4–8

If in a field theory we have

$$[\varphi(x), \psi(y)]_{\pm} = 0 \qquad \text{for } (x-y)^2 < 0,$$

and the opposite (4–37)

$$[\varphi(x), \psi^*(y)]_{\mp} = 0$$

then either φ or ψ vanishes.

Proof:

If f and g are two test functions of compact support we have

$$(\Psi_0, \varphi(f)^*\psi(g)^*\psi(g)\varphi(f)\Psi_0) = \|\psi(g)\varphi(f)\Psi_0\|^2 \geqslant 0. \tag{4–38}$$

If the supports of f and g are space-like separated, the assumed commutation relations (4–37) imply that the left-hand side of (4–38) is

$$-(\Psi_0, \psi(g)^*\psi(g)\varphi(f)^*\varphi(f)\Psi_0). \tag{4–39}$$

By the cluster decomposition property, if the support of g runs to infinity in a space-like direction, (4–39) approaches

$$-(\Psi_0, \psi(g)^*\psi(g)\Psi_0)(\Psi_0, \varphi(f)^*\varphi(f)\Psi_0) = -\|\psi(g)\Psi_0\|^2\|\varphi(f)\Psi_0\|^2$$

and this is not positive. Therefore, comparing with (4–38) we see that 0 is the only consistent value of the limit, and either $\psi(g)\Psi_0 = 0$ or $\varphi(f)\Psi_0 = 0$. By Theorem 4–3 this implies either $\psi(g) = 0$ or $\varphi(f) = 0$. If $\psi \neq 0$ there exists a test function of compact support g for which $\psi(g) \neq 0$. Then, for all test functions f of compact support, $\varphi(f) = 0$. Since the test functions of compact support are dense in $\mathscr{S}$, this implies $\varphi = 0$. Similarly, if $\varphi \neq 0$, then $\psi = 0$. ▮

Corollary

In a field theory there can be no non-zero field satisfying

$$[\varphi(x), \varphi(y)]_{\pm} = 0 \tag{4–40}$$

and

$$[\varphi(x), \varphi^*(y)]_{\mp} = 0, \quad \text{for all } (x - y)^2 < 0.$$

Next we prove the proper spin-statistics theorem. For clarity, we first deal with a scalar field and then describe the modifications necessary for arbitrary spin.

Theorem 4–9 (*Spin-Statistics Theorem for a Scalar Field*)

Let φ be a scalar field. Suppose that

$$[\varphi(x), \varphi^*(y)]_+ = 0 \qquad \text{for } (x - y)^2 < 0. \tag{4–41}$$

Then $\varphi(x)\Psi_0 = 0 = \varphi(x)^*\Psi_0$. In a field theory in which φ and φ^* commute or anti-commute with all other fields this implies $\varphi = \varphi^* = 0$.

Proof:

The hypothesis (4–41) of the "wrong" connection of spin with statistics implies, if $(x - y)^2 < 0$,

$$(\Psi_0, \varphi(x)\varphi^*(y)\Psi_0) + (\Psi_0, \varphi^*(y)\varphi(x)\Psi_0) = 0. \tag{4–42}$$

Each of the vacuum expectation values in this equation depends only on $(x - y)$, and is the boundary value of a holomorphic function

$$
\begin{aligned}
(\Psi_0, \varphi(x)\varphi^*(y)\Psi_0) &= \lim_{\substack{\eta\to 0\\ \eta\in V_+}} W(x - y - i\eta)\\
(\Psi_0, \varphi^*(y)\varphi(x)\Psi_0) &= \lim_{\substack{\eta\to 0\\ \eta\in V_+}} \hat{W}(y - x - i\eta)
\end{aligned}
\tag{4–43}
$$

where W and $\hat{W}$ are holomorphic for $\eta \in V_+$. Because W and $\hat{W}$ are invariant under the restricted Lorentz group, Theorem 2–11 implies they are invariant under the proper complex Lorentz group, and are holomorphic in the extended tube $\mathscr{T}_1'$ in the variable $\zeta = (x - y) - i\eta$, which includes all points where ζ is real and space-like. Thus, (4–42) is a relation between holomorphic functions

$$W(\zeta) + \hat{W}(-\zeta) = 0, \tag{4–44}$$

which holds at an open set of real points of the extended tube and therefore throughout the extended tube. Using the invariance of $\hat{W}$ under the complex proper Lorentz transformation $\Lambda = -1$,

$$\hat{W}(\zeta) = \hat{W}(-\zeta), \tag{4–45}$$

and we get from (4–44)

$$W(\zeta) + \hat{W}(\zeta) = 0, \tag{4–46}$$

valid for $\zeta \in \mathscr{T}_1'$.

Passing to the boundary values in (4–46) as $\eta \to 0$, $\eta \in V_+$, we obtain for *all* x and y the distribution relation

$$(\Psi_0, \varphi(x)\varphi^*(y)\Psi_0) + (\Psi_0, \varphi^*(-y)\varphi(-x)\Psi_0) = 0. \tag{4–47}$$

We claim that this equation (4–47) implies $\varphi(x)\Psi_0 = 0$. To prove this, set $\hat{f}(x) = f(-x)$. Then, recalling that $\varphi^*(f) = [\varphi(\bar{f})]^*$,

$$
\varphi(f) = \int dx\, f(x)\varphi(x), \quad \text{and} \quad
\begin{aligned}[t]
\varphi(\hat{f}) &= \int dx\, f(-x)\varphi(x)\\
&= \int dx\, f(x)\varphi(-x),
\end{aligned}
$$

we get

$$
\begin{aligned}
\|\varphi(f)^*\Psi_0\|^2 + \|\varphi(\hat{f})\Psi_0\|^2 &= (\Psi_0, \varphi(f)\varphi(f)^*\Psi_0) + (\Psi_0, \varphi(\hat{f})^*\varphi(\hat{f})\Psi_0)\\
&= \int dx\, dy\, f(x)\overline{f(y)}[(\Psi_0, \varphi(x)\varphi^*(y)\Psi_0)\\
&\qquad + (\Psi_0, \varphi^*(-y)\varphi(-x)\Psi_0)]\\
&= 0.
\end{aligned}
$$

Thus, for all test functions, f, $\|\varphi(f)\Psi_0\| = \|\varphi(f)^*\Psi_0\| = 0$, which yields $\varphi(f)\Psi_0 = 0$ and completes the proof of the first half of the theorem.

In any field theory in which all the fields commute or anti-commute, Theorem 4–3 applies, and so $\varphi(f)\Psi_0 = 0$ implies $\varphi = \varphi^* = 0$. ▌

For fields of general spin an analogue of Theorem 4–9 is

Theorem 4–10 (*Spin-Statistics Theorem for General Spin*)

For a general irreducible spinor field the "wrong" connection of spin with statistics

$$[\varphi_\alpha(x), \varphi_\alpha^*(y)]_+ = 0 \quad \varphi \text{ of integer spin}$$
$$[\varphi_\alpha(x), \varphi_\alpha^*(y)]_- = 0 \quad \varphi \text{ of half-odd integer spin}$$
$$\text{for } (x-y)^2 < 0 \qquad (4\text{–}48)$$

implies $\varphi_\alpha(x)\Psi_0 = 0$. In a field theory in which all fields either commute or anti-commute this implies $\varphi_\alpha = \varphi_\alpha^* = 0$.

Proof:

Without confusion we may leave out the suffix α. The proof follows the same lines as for a scalar field. As the generalization of (4–42), the relations (4–48) give directly

$$(\Psi_0, \varphi(x)\varphi^*(y)\Psi_0) \pm (\Psi_0, \varphi^*(y)\varphi(x)\Psi_0) = 0, \quad \text{for } (x-y)^2 < 0, \quad (4\text{–}49)$$

the sign being the same as in (4–48). As in the proof of Theorem 4–9, there are holomorphic functions W and $\hat{W}$ such that the vacuum expectation values are given by (4–43). The generalization of (4–44) is

$$W(\zeta) \pm \hat{W}(-\zeta) = 0, \qquad (4\text{–}50)$$

while that of (4–45) is

$$\hat{W}(\zeta) = (-1)^J \hat{W}(-\zeta), \qquad (4\text{–}51)$$

where J is the total number of undotted indices in $\varphi\varphi^*$. Equation (4–51) is a consequence of the transformation law of $\hat{W}$ under the group $SL(2, C) \otimes SL(2, C)$ and is a special case of (4–32). We remark that the number of undotted indices in $\varphi\varphi^*$ is the same as the total number of indices in φ, and so is even (odd) if the field φ is of integer (half-odd integer) spin. Therefore, at points of holomorphy,

$$\hat{W}(\zeta) = \pm \hat{W}(-\zeta), \qquad (4\text{–}52)$$

the sign being the same as in (4–48). Combining (4–52) and (4–50) we arrive at (4–46), in exactly the same form as before. From here on, the argument is the same as for Theorem 4–9. ▌

Now we turn to a discussion of the commutation relations between different fields. We begin with an example, which illustrates a number of the principles involved in a simple way.

Example 1: Anti-commuting Scalar and Spin ½ Field

Suppose φ is a scalar field and ψ is a spin ½ field, and that φ and ψ anti-commute at space-like separations. We can define new fields φ' and ψ' by

$$\varphi'(x)\Psi_1 = \varphi(x)\Psi_1, \qquad \psi'(x)\Psi_1 = \psi(x)\Psi_1$$

if Ψ_1 is a vector in the domain of definition D of φ,ψ in the coherent subspace $\mathscr{H}_1$ of $\mathscr{H}$ corresponding to an integer spin representation of $\mathscr{P}_+^\uparrow$, and

$$\varphi'(x)\Psi_2 = -\varphi(x)\Psi_2, \qquad \psi'(x)\Psi_2 = \psi(x)\Psi_2$$

if Ψ_2 is a vector in D in the coherent subspace $\mathscr{H}_2$ of $\mathscr{H}$ corresponding to a half-odd integer spin representation of $\mathscr{P}_+^\uparrow$. The action of φ', ψ' on linear superpositions of states from $\mathscr{H}_1$ and $\mathscr{H}_2$ is given by linearity. A super-selection rule (the univalence super-selection rule) operates between $\mathscr{H}_1$ and $\mathscr{H}_2$. Recall from Section 1–1 that linear superpositions of states from different coherent subspaces between which there is a super-selection rule are not physically realizable. In this case, a state $\alpha\Psi_1 + \beta\Psi_2$ becomes $\alpha\Psi_1 - \beta\Psi_2$ when it is rotated through an angle of 2π around any axis. This should belong to the same ray, and so that state is not physically realizable unless $\alpha = 0$ or $\beta = 0$. It is easy to see that the new fields, φ', ψ' commute at space-like separated points, that is, in contrast to φ,ψ the new fields satisfy the normal commutation relations. The transformation from φ, ψ to φ', ψ' is called a Klein transformation. Unlike the case of the symmetries we have discussed in Section 3–5, there is no unitary or anti-unitary transformation V such that

$$\varphi' = V\varphi V^{-1}, \qquad \psi' = V\psi V^{-1}$$

because the existence of such a V would imply

$$[\varphi'(x), \psi'(y)]_- = V[\varphi(x), \psi(y)]_- V^{-1}.$$

We can take two points of view with respect to the transformation $\varphi, \psi \to \varphi', \psi'$. In the first, it is simply "a change of variables." The observables of the theory will be certain functions of the φ, ψ, and φ, ψ are functions of the φ', ψ'. Hence the observables are certain (different) functions of the fields ψ', φ'. The preceding discussion merely says that by making a change of variables one can express the observables in terms of a set of fields having normal commutation relations.

In the second point of view, one makes a different correspondence between operators and laboratory operations. If a certain function, say $F(\varphi, \psi)$, is observable and corresponds to certain well-defined measurements in the laboratory, then the Klein transformation is regarded as leading to a new theory in which $F(\varphi', \psi')$ corresponds to that very set of measurements. It is natural to ask whether the two theories predict the same result for all experiments. Whether this is so only a detailed analysis of the measurement theory of the system can decide. The more operators that are required to be observable, the less chance there is that the two theories are physically equivalent. In our example we have

$$(\Psi_2, \varphi(x)\Psi_2) = -(\Psi_2, \varphi'(x)\Psi_2),$$

so if $\varphi(x)$ is observable then the two theories are different. At the other extreme, where only the results of scattering experiments are observable, the theories predict the same results; this is discussed below in connection with the *PCT* theorem.

The example just considered has one feature not characteristic of the general situation for abnormal commutation relations; the existence of the orthogonal subspaces $\mathscr{H}_1$ and $\mathscr{H}_2$ invariant under $U(a, \Lambda)$ is already a consequence of the univalence super-selection rule. In general, the existence, orthogonality, and invariance of the analogous subspaces is a consequence of the abnormal commutation relations themselves. Since several new points of principle arise in this situation, we illustrate it by means of a second simple example.

Example 2: A Scalar and Two Spin ½ Fields

Consider a theory of one hermitian scalar field φ and two hermitian half-odd integer spin fields ψ_1 and ψ_2. For simplicity the spinor indices on ψ_1 and ψ_2 will be suppressed. Suppose the normal commutation relations

$$[\psi_1(x), \psi_2(y)]_+ = 0 = [\varphi(x), \psi_2(y)]_- \qquad (x - y)^2 < 0 \tag{4–53}$$

hold, together with the "abnormal" relation

$$[\varphi(x), \psi_1(y)]_+ = 0, \qquad (x - y)^2 < 0. \tag{4–54}$$

These relations imply the vanishing of certain vacuum expectation values, as will now be shown. For later application the result will be stated and proved for general field theories.

Theorem 4–11

In any local field theory, if $M(x_1, \ldots x_j)$ and $N(y_1, \ldots y_k)$ are two monomials in the field components which anti-commute if (x) and (y) are space-like separated sets of points, then either $(\Psi_0, M\Psi_0) = 0$, or $(\Psi_0, N\Psi_0) = 0$.

Proof:

Consider

$$(\Psi_0, M(x_1, \ldots x_j)N(y_1 + a, \ldots y_k + a)\Psi_0) = -(\Psi_0, N(y_1 + a, \ldots y_k + a)M(x_1, \ldots x_j)\Psi_0) \quad (4\text{–}55)$$

if a is a large space-like vector. The limit of this equation as $a \to \infty$ is, by the cluster decomposition theorem (Theorem 3–4),

$$(\Psi_0, M\Psi_0)(\Psi_0, N\Psi_0) = -(\Psi_0, N\Psi_0)(\Psi_0, M\Psi_0).$$

Therefore, one or the other of $(\Psi_0, M\Psi_0)$ and $(\Psi_0, N\Psi_0)$ must vanish. ▮

Returning again to the model, let us suppose that for some odd n, $(\Psi_0, \varphi(x_1)\ldots\varphi(x_n)\Psi_0) \neq 0$. The alternative possibility can, of course, also be analyzed, but this case suffices to illustrate our point. Putting

$$M = \varphi(x_1)\ldots\varphi(x_n) \qquad n \text{ odd}$$

$$N = \varphi(y_1)\ldots\varphi(y_j)\psi_1(y_{j+1})\ldots\psi_1(y_{j+k})\psi_2(y_{j+k+1})\ldots\psi_2(y_{j+k+\ell}) \qquad k \text{ odd}, \ \ell \text{ odd}$$

and applying Theorem (4–11), since M and N anti-commute we get for all j

$$(\Psi_0, \varphi(y_1)\ldots\varphi(y_j)\psi_1(y_{j+1})\ldots\psi_1(y_{j+k})\psi_2(y_{j+k+1})\ldots\psi_2(y_{j+k+\ell})\Psi_0) = 0 \qquad k \text{ odd}, \ \ell \text{ odd}. \quad (4\text{–}56)$$

The expression already vanishes if $k + \ell$ is odd [see the remark after (4–30)]. Clearly this and (4–56) together imply that the vacuum expectation values, and therefore the theory, are invariant under the group of four transformations given by $(\psi_1, \psi_2) \to (\pm\psi_1, \pm\psi_2)$. This symmetry gives rise to a conservation law, which will be called an *even-odd rule*.

The Hilbert space $\mathscr{H}$ splits into a sum of four orthogonal subspaces $\mathscr{H}_{\pm,\pm}$, spanned respectively by states

$$\varphi(x_1)\ldots\varphi(x_j)\psi_1(x_{j+1})\ldots\psi_1(x_{j+k})\psi_2(x_{j+k+1})\ldots\psi_2(x_{j+k+\ell})\Psi_0$$

with (k, ℓ) = (even, even), (even, odd), (odd, even), and (odd, odd). The orthogonality of the four subspaces is guaranteed by (4–56) together with the univalence super-selection rule. Clearly, they are invariant under the transformations of the Poincaré group.

We now define the Klein transformation for this case. It leaves ψ_1 and ψ_2 invariant and replaces φ by a new field φ', defined as φ on those vectors of the domain of φ which are in $\mathscr{H}_{+,\pm}$, $-\varphi$ on those which are in $\mathscr{H}_{-,\pm}$, and by linearity elsewhere. It can be seen that φ' is hermitian. Then φ', together with $\psi_1' = \psi_1$ and $\psi_2' = \psi_2$, satisfy the normal commutation relations.

Analogues of the remarks made on the first example apply here. In addition, there are the following new features. The even-odd rule defines a conservation law for the system; does it define a superselection rule? Again, an analysis of the observables is necessary to answer such questions. The new theory is equivalent to the old in the first sense discussed above: there exists an irreducible set of field operators $(\varphi', \psi_1', \psi_2')$ which have normal commutation relations. It may be, of course, that interesting quantities of the theory are more "simply" expressible in terms of one set of fields than the other. In this connection there is a good reason for preferring the transformed fields $(\varphi', \psi_1', \psi_2')$, for the hermitian quantities $\varphi(x)$ and $\psi_1(y)\psi_2(z) + \psi_2(z)\psi_1(y)$ do not commute at large space-like distances. This seems unnatural in a field theory; it means that any function of the fields $(\varphi, \psi_1, \psi_2)$ which corresponds to a local measurement of the system must be a rather complicated function of the old fields. It might be a simple function of the new fields, which are not beset by lack of commutativity at space-like separations. Our discussion of the second example is now complete.

We now deal with the case of a general set of local fields, following the treatment given by Araki. The essential steps are to show that abnormal commutation relations give rise to even-odd rules and then to show that this makes possible the definition of the required Klein transformations. We shall suppose that there are n fields $\varphi_1 \ldots \varphi_n$ and that all components of φ_j satisfy the same commutation (or anti-commutation) relations with all components of φ_k. We shall assume the adjoints $\varphi_{j\alpha}^*$ of the components $\varphi_{j\alpha}$ of φ_j occur somewhere in the list of $\varphi_{k\beta}$. Our ultimate objective is given in the following theorem.

Theorem 4–12

In any field theory with abnormal commutation relations there always exists an irreducible set of fields with normal commutation relations obtained from the original set by a Klein transformation.

The proof is rather involved, and will be carried out in a number of stages. The first step is to generalize the notion of an even-odd rule. A theory is said to have an *even-odd rule for a set* α of fields if every vacuum expectation value with an odd number of fields from α vanishes. In such a case the two spaces $\mathscr{H}_o$ and $\mathscr{H}_e$ generated from the vacuum by polynomials which are odd or even, respectively, in the fields belonging to α (and arbitrary in the rest of the fields), are orthogonal to each other, and are obviously invariant under $\mathscr{P}_+^\uparrow$. Hence the operator $p(\alpha)$, defined to be 1 on $\mathscr{H}_e$ and -1 on $\mathscr{H}_o$, and by linearity elsewhere, commutes with the representation $U(a,\Lambda)$ of $\mathscr{P}_+^\uparrow$.

Suppose we define a Klein transformation to new fields by multiplying all the fields in another subset β of the fields by $p(\alpha)$:

$$\begin{aligned} \varphi_j' &= p(\alpha)\varphi_j, \qquad \varphi_j \in \beta \\ \varphi_j' &= \varphi_j, \qquad \varphi_j \notin \beta \end{aligned} \tag{4–57}$$

Then φ' again are fields, in general with different commutation relations. The set α in (4–57) (or several such sets) is determined by the abnormal commutation relations, i.e., by applying Theorem 4–11. The set β is free to be chosen appropriately so that the commutation relations are changed to the normal ones. In the first example above, α is the set consisting of the spin $\frac{1}{2}$ field ψ, and the even-odd rule is that given by the univalence super-selection rule. In the second example there are two possible α's, the set consisting of ψ_1 or the set consisting of ψ_2. Our choice of the set β was, in both examples, the set consisting of the single field φ.

The commutation relations of φ_j', φ_k' are different from those of φ_j, φ_k if one field is in α and the other is in β, and both fields are not in both sets. Otherwise the relations are not altered. To see this we must consider in turn all the possible cases, of which there are sixteen, enumerated according as φ_1 or φ_2 is or is not in the set α or β. (For simplicity, j and k are replaced by 1 and 2 temporarily.) If neither φ_1 nor φ_2 lies in the set β, then $\varphi_1' = \varphi_1$ and $\varphi_2' = \varphi_2$, and so the relations are unchanged. This takes care of four cases, labeled by ($\varphi_1 \in \alpha, \notin \beta$; $\varphi_2 \in \alpha, \notin \beta$); ($\varphi_1 \notin \alpha, \notin \beta$; $\varphi_2 \in \alpha, \notin \beta$); ($\varphi_1 \in \alpha, \notin \beta$; $\varphi_2 \notin \alpha, \notin \beta$); ($\varphi_1 \notin \alpha, \notin \beta$; $\varphi_2 \notin a, \notin \beta$). If neither φ_1 nor φ_2 is in α then both fields commute with $p(\alpha)$, and the commutation rule is unchanged. This accounts for three more cases: ($\varphi_1 \notin \alpha, \notin \beta$; $\varphi_2 \notin \alpha, \in \beta$); ($\varphi_1 \notin \alpha, \in \beta$; $\varphi_2 \notin \alpha, \in \beta$); and ($\varphi_1 \notin \alpha, \in \beta$; $\varphi_2 \notin \alpha, \notin \beta$). It is also easy to see that if one of the fields is in neither set, the Klein transformation does not alter the commutator. This accounts for ($\varphi_1 \notin \alpha, \notin \beta$; $\varphi_2 \in \alpha, \in \beta$); ($\varphi_1 \in \alpha, \in \beta$; $\varphi_2 \notin \alpha, \notin \beta$). If both φ_1 and φ_2 lie in both sets, then $\varphi_1'\varphi_2' = p(\alpha)\varphi_1 p(\alpha)\varphi_2 = -\, p(\alpha)^2\varphi_1\varphi_2 = -\varphi_1\varphi_2$ and $\varphi_2'\varphi_1' = -\varphi_2\varphi_1$. Therefore the vanishing of $[\varphi_1, \varphi_2]_\pm$ is unchanged by (4–57). This disposes of ($\varphi_1 \in \alpha, \in \beta$; $\varphi_2 \in \alpha, \in \beta$). The remaining six possibilities satisfy the condition that one of φ_1, φ_2 lies in α and the other in β. They are ($\varphi_1 \notin \alpha, \in \beta$; $\varphi_2 \in \alpha, \in \beta$); ($\varphi_1 \in \alpha, \notin \beta$; $\varphi_2 \notin \alpha, \in \beta$); ($\varphi_1 \in \alpha, \notin \beta$; $\varphi_2 \in \alpha, \in \beta$); ($\varphi_1 \in \alpha, \in \beta$; $\varphi_2 \notin \alpha, \in \beta$); ($\varphi_1 \in \alpha, \in \beta$; $\varphi_2 \in \alpha, \notin \beta$); ($\varphi_1 \notin \alpha$,

$\in \beta$; $\varphi_2 \in \alpha, \notin \beta$). It is left to the reader to check that in these six cases, $[\varphi_1, \varphi_2]_\pm = 0$ changes into $[\varphi_1', \varphi_2']_\mp = 0$, using (4–57) and the fact that if $\varphi_j \in \alpha$, $p(\alpha)$ anti-commutes with φ_j, whereas if $\varphi_j \notin \alpha$, $p(\alpha)$ and φ_j commute. We therefore have the result: the transformation (4–57) alters the commutation relations if one field is in α and the other is in β and both fields are not in both sets; otherwise the relations are unchanged.

We want to transform the commutation relations to the normal ones, and so the matrix σ_{ij} defined as follows is of interest:

$$\begin{aligned} \sigma_{ij} &= 0 \quad \text{if } \varphi_i, \varphi_j \text{ have normal commutation relations} \\ \sigma_{ij} &= 1 \quad \text{otherwise.} \end{aligned} \tag{4–58}$$

Obviously $\sigma_{ij} = \sigma_{ji}$; Theorem 4–10 asserts that a field always has normal commutation relations with itself, which means that $\sigma_{ii} = 0$.

In order to determine whether we have an even-odd rule we examine the commutation relations of monomials $M, N, \ldots$ of the fields, and we can introduce the concept of *normal commutation relations between monomials*: two monomials M and N are said to have "normal" commutation relations if they commute (anti-commute) just as if all the fields in them had normal commutation relations. Otherwise the monomials are said to have "abnormal" commutation relations. Two monomials may have normal commutation relations although their constituent fields have not. It is clear that the commutation relations of one monomial with another depends only on the parities of the powers of each field in them.† In other words, two monomials, M_1 and M_2, have the same relations with any other monomial if M_1 and M_2 belong to the same equivalence class defined as follows: *two monomials are equivalent if the number of times each field φ_i occurs has the same parity in both.* Each equivalence class contains a monomial of least degree, in which a given field φ_j either occurs once or not at all (the order in which the fields are multiplied does not matter for our purposes). Thus, there is a 1:1 correspondence between the equivalence classes and subsets of the fields $\varphi_1, \ldots \varphi_n$. We shall label each equivalence class by a vector $s = (s_1, \ldots s_n)$ defined as follows:

$$s_j = \begin{pmatrix} 0 \\ 1 \end{pmatrix} \text{ if } \varphi_j \text{ appears an } \begin{pmatrix} \text{even} \\ \text{odd} \end{pmatrix} \text{ number of times}$$

in the monomials in the equivalence class s. If M is a monomial we will use the symbol $s(M)$ to denote the equivalence class containing

† Two numbers are said to have the same *parity* if they are both even, or both odd.

M or the vector which labels it. To any set β of fields we can associate the vector $t(\beta) = (t_1(\beta),\ldots t_n(\beta))$, which is defined by

$$t_j(\beta) = 0 \qquad \text{if } \varphi_j \notin \beta$$
$$t_j(\beta) = 1 \qquad \text{if } \varphi_j \in \beta.$$

If β is the set $(\varphi_1,\ldots\varphi_k)$, then clearly $t(\beta)$ is the equivalence class containing the monomial $\varphi_1\ldots\varphi_k$.

We shall define addition of two vectors s,t by

$$(s + t)_j = s_j + t_j \ (\text{mod } 2),$$

that is, we use the addition rule for components $0 + 0 = 1 + 1 = 0$; $0 + 1 = 1 + 0 = 1$. This is convenient, since we are only interested in the parity of the numbers involved. With this addition law the equivalence classes $s,\ldots$ form a vector space V. It is easy to derive the following two formulas:

$$\sum_j s_j(M)t_j(\beta) = \begin{pmatrix}0\\1\end{pmatrix} \ (\text{mod } 2) \tag{4–59}$$

if M contains an $\begin{pmatrix}\text{even}\\\text{odd}\end{pmatrix}$ number of fields from β, and from (4–58),

$$\sum_{i,j} s_i(M)\sigma_{ij}s_j(N) = \begin{pmatrix}0 & \text{if } M,N \text{ obey the normal commutation relations}\\1 & \text{otherwise,}\end{pmatrix} \tag{4–60}$$

the sum again being mod 2.

We now have most of the apparatus necessary to prove the theorem. The effect of the transformation (4–57) on σ_{ij} can be calculated explicitly. If i and j are such that φ_i is in one of the sets α,β, and φ_j is in the other, then from the definition of $t(\alpha)$, $t(\beta)$

$$\begin{aligned}\text{either}\quad & t_i(\alpha) = t_j(\beta) = 1\\ \text{or}\quad & t_i(\beta) = t_j(\alpha) = 1.\end{aligned} \tag{4–61}$$

We have remarked that σ_{ij} is changed by the Klein transformation (4–57) if and only if one and only one of (4–61) holds. Thus the effect of (4–57) on σ is

$$\sigma_{ij} \rightarrow \sigma_{ij} + t_i(\alpha)t_j(\beta) + t_j(\alpha)t_i(\beta) \ (\text{mod } 2). \tag{4–62}$$

We show that σ can be reduced to 0 by a series of transformations (4–57) by proving that σ has the form given by the following theorem.

Theorem 4–13

If $\sigma_{ij} = \begin{pmatrix} 0 \\ 1 \end{pmatrix}$ if φ_i and φ_j have $\begin{pmatrix} \text{normal} \\ \text{abnormal} \end{pmatrix}$ commutation relations, then

$$\sigma_{ij} = \sum_{k=1}^{N} [t_i(\alpha_k)s_j^{(k)} + s_i^{(k)}t_j(\alpha_k)] \text{ (mod 2)}, \qquad (4\text{–}63)$$

where the α_k are sets having an even-odd rule, and the $s^{(k)}$, $k = 1, \ldots N$ are some vectors in V.

Proof of Theorem 4–12:

Assuming the result of Theorem 4–13 we can proceed as follows. Choose the set β_k to consist of the fields in the minimal monomial in $s^{(k)}$ of minimal degree in each of the fields [i.e., $s^{(k)} = t(\beta_k)$]. If we apply (4–57) N times using the α_k, β_k, then σ_{ij} is transformed to

$$[\sigma_{ij} + \sigma_{ij}] \text{ (mod 2)} = 0 \text{ (mod 2)}. \quad \blacksquare$$

Proof of Theorem 4–13:

To prove Theorem 4–13 we must have a criterion for deciding whether a given set α has an even-odd law. For this we introduce the subset Γ of V defined by: $s \in \Gamma$ if there exists a monomial M in the equivalence class s such that $(\Psi_0, M\Psi_0) \neq 0$. We can define $\bar{\Gamma}$ to be the set obtained by adjoining to Γ all sums of vectors in Γ. [For example, if Γ contains (0,1) and (1,1), then $\bar{\Gamma}$ contains (0,1), (1,1), and (0,1) + (1,1) = (1,0).] The set Γ (or $\bar{\Gamma}$) is just what is wanted in order to determine the even-odd rules of the theory, as will now be explained.

Our assertion is the following: if

$$\sum_j s_j t_j(\alpha) = 0 \text{ (mod 2)} \quad \text{for all } s \in \Gamma \qquad (4\text{–}64)$$

(and incidentally, therefore, for all $s \in \bar{\Gamma}$), then there is an even-odd rule for the set α. For the proof, suppose, if possible, that M is a monomial containing an odd number of fields from α (and any number of other fields) and is such that $(\Psi_0, M\Psi_0) \neq 0$. Let $s(M)$ be the equivalence class of M. By the definition of Γ, $s(M) \in \Gamma$, and by (4–59), $\sum_j s_j(M)t_j(\alpha) = 1$ (mod 2), contradicting the given property.

We next assert that if $s(M)$, $s(N)$ are two vectors in Γ, then M,N have normal commutation relations. For $(\Psi_0, M\Psi_0) \neq 0$, $(\Psi_0, N\Psi_0) \neq 0$ and therefore M and N commute to avoid contradicting Theorem 4–11. Both M and N contain an even number of half-odd integer spin fields and so, if we had assumed that all fields φ_j had normal relations, M and N would commute. Thus, M and N have normal commutation relations. Equivalent to this statement, according to (4–60), is the condition

$$\sum_{i,j} s(M)_i \sigma_{ij} s(N)_j = 0 \qquad \text{if } s(M),\, s(N) \in \Gamma, \tag{4–65}$$

which will play a significant role in the following. This concludes the preliminary discussion, and we now turn to the proof proper.

Let $e^{(1)}, \ldots e^{(m)}$ be a linearly independent set of vectors which span $\overline{\Gamma}$, and $e^{(1)}, \ldots e^{(n)}$ a basis in V which includes $e^{(1)}, \ldots e^{(m)}$.

We can always find a (dual) linearly independent set $d^{(k)}$ such that

$$(e^{(j)}, d^{(k)}) = \sum_i e_i^{(j)} d_i^{(k)} = \delta_{jk} \ (\text{mod } 2). \tag{4–66}$$

We are here using the natural scalar product in V:

$$(s, t) = \sum_{i=1}^{n} s_i t_i \ (\text{mod } 2).$$

Note that this scalar product is not positive definite, so that mutual orthogonality for a set of vectors does not imply linear independence. For example, $((1,1), (1,1)) = 1 + 1 = 0$ and so $(1,1)$ is orthogonal to itself. However, this lack of positive definiteness does not prevent us from finding the $d^{(k)}$.

We can regard σ as an operator in V: $s \to \sigma s$, given by

$$(\sigma s)_j = \sum_i \sigma_{ij} s_i \ (\text{mod } 2).$$

Let us denote the matrix elements of σ in the new coordinate system $e^{(1)}, \ldots e^{(n)}$ by

$$\sigma'_{ij} = (e^{(i)}, \sigma e^{(j)}). \tag{4–67}$$

Then

$$(t, \sigma s) = \sum_{i,k} (t, d^{(i)}) \sigma'_{ik} (d^{(k)}, s); \qquad s, t, \in \text{V}. \tag{4–68}$$

We know that any monomial has normal relations with itself. This applies to the monomials defined by the vectors $e^{(i)}$, and so, using (4–60) we see that $\sigma'_{ii} = 0$, $i = 1, 2, \ldots n$. From the definition we easily see that

$$\sigma'_{ij} = \sigma'_{ji}.$$

Because the vectors $e^{(1)},\dots e^{(m)}$ lie in Γ, $\sigma'_{ik} = 0$ if $i,k \leqslant m$, as a consequence of (4–65). Therefore

$$\sigma_{ij} = \sum_{k>m} (d_i^{(k)}s_j^{(k)} + s_i^{(k)}d_j^{(k)}),$$

where

$$s^{(k)} = \sum_{i<k} \sigma'_{ik}d^{(i)}.$$

We now assert that the vectors $d^{(k)}$ define sets α_k with an even-odd rule, if $k > m$. Let α_k consist of those fields present in the monomial in the equivalence class defined by $d^{(k)}$ of minimal degree in each of the fields, that is,

$$d^{(k)} = t(\alpha_k), \quad \text{or} \quad \varphi_j \in \alpha_k \quad \text{if } d_j^{(k)} = 1$$
$$\varphi_j \notin \alpha_k \quad \text{if } d_j^{(k)} = 0.$$

Then if $k > m$ we have $(d^{(k)},e^j) = 0$ for $j \leqslant m$, which gives $\sum_i d_i^{(k)}s_i = 0$ (mod 2) for all $s \in \Gamma$, since e^j, $j = 1,\dots m$, span Γ. Therefore, by (4–64), the set α_k defines an even-odd rule. ∎

We now return to the discussion of the *PCT* theorem, proved in the last section to hold in any theory with the normal commutation relations. What is the situation when the commutation relations are abnormal? The easiest way to answer this question is to use the existence of the Klein transformation of Theorem 4–12 from the given fields φ_i to fields φ'_i satisfying normal commutation relations.

Applying the *PCT* theorem to the new fields φ'_i we see that there exists a symmetry Θ in the theory which satisfies

$$\Theta\varphi'_i(f)\Theta^{-1} = (-1)^j\binom{i}{1}\varphi_i^{*\prime}(\hat{f}) \tag{4–69}$$
$$\Theta\Psi_0 = \Psi_0$$

the upper case for half-odd integer spin fields, the lower case for integer spin fields. In (4–69), $\hat{f}(x) = \bar{f}(-x)$. This will induce a transformation of the original fields φ_i which will, in general, *not* be of the form (4–69) with the primes removed. To see how this comes about consider a third example.

Example 3: Anti-Commuting Hermitian Scalar Fields

Let φ and ψ be two anti-commuting hermitian scalar fields such that

$$(\Psi_0, \varphi(x)\psi(y)\Psi_0) \neq 0.$$

We conclude from Theorem 4–11 that all vacuum expectation values containing an odd number of fields vanish. We can define a Klein transformation for this case by $\varphi' = \varphi$, $\psi' = \psi$ on the subspace spanned by even polynomials in the smeared fields applied to the vacuum and

$\varphi' = -\varphi$, $\psi' = \psi$ on the subspace spanned by the odd polynomials applied to the vacuum. Then

$$(\varphi\psi\Psi_0, \Psi_0) = (\psi\Psi_0, \varphi\Psi_0), \quad \text{i.e.,} \quad -(\varphi'\psi'\Psi_0, \Psi_0) = (\psi'\Psi_0, \varphi'\Psi_0),$$

so $(\varphi')^* = -\varphi'$; a Klein transformation may take a hermitian field into a non-hermitian field.

Thus, in general, we will get $(\varphi_i')^* = \pm (\varphi_i{}^*)'$, and combining this with (4–69) we arrive at a transformation law for the fields φ_i which differs from the usual one by a change of sign for some of the fields. The interpretation of Θ as the *PCT* operator can be established just as for the normal case given at the end of the last section. It remains to show that the *particles* of the theory obey the "normal" statistics. This can be done using the asymptotic condition proved in Ref. 6. A state containing several incoming particles can be obtained by taking the strong limit, as $t \to -\infty$, of a state Ψ_t labeled by a time parameter. The state Ψ_t may be created from the vacuum by applying a monomial in the Klein-transformed fields. Since these fields obey normal commutation relations, Ψ_t satisfies normal statistics at finite times, and therefore asymptotically.

4–5. HAAG'S THEOREM AND ITS GENERALIZATIONS

We have already remarked in Section 3–1 that in the traditional approach to field theory one assumed that the operators of an irreducible set satisfied the canonical commutation relations at a given time. In this section we follow tradition and pursue this idea further.

It can be proved for systems with a finite number of degrees of freedom that under certain conditions any two solutions of the commutation relations are connected by a canonical (unitary) transformation. If this result were taken over to field theory we could define the so-called *interaction picture*. The canonical variables at each time are assumed to be equivalent to the canonical variables of a free field φ_{int}. In particular,

$$V(t)\varphi(\mathbf{x}, t)V(t)^{-1} = \varphi_{\text{int}}(\mathbf{x}, t). \tag{4–70}$$

The time dependence of the operator V reflects the presence of interactions, and various interesting quantities can be calculated in terms of it. For example, the collision operator S is given by

$$S = \lim_{t\to\infty} V(t)V(-t)^*.$$

It was soon realized that the argument leading to (4–70) is inconclusive because there are many inequivalent representations of the canonical commutation relations [that is, representations which cannot be connected by a unitary transformation (4–70)]. However, the argument is not only inconclusive, it is *wrong*: unless $\varphi(\mathbf{x},t)$ is a free field there is no V satisfying (4–70), as will now be shown by arguments which realize ideas of R. Haag.

Evidently, in order for this theorem to make sense we have to assume fields which are distributions in their space arguments for each value of their time arguments.

Our first step is to show that the unitary equivalence of two irreducible sets of operators implies the equality of their equal time vacuum expectation values.

Theorem 4–14

Let $\varphi_{1\alpha}(f, t)$ and $\varphi_{2\beta}(f, t)$, $f \in \mathscr{S}(\mathbf{R}^3)$ be any two irreducible sets of field operators at time t defined respectively in Hilbert spaces $\mathscr{H}_1$ and $\mathscr{H}_2$, in which there are continuous unitary representations of the inhomogeneous SU_2 group,

$$\{\mathbf{a}, A\} \rightarrow U_j(\mathbf{a}, A) \qquad j = 1, 2,$$

respectively, such that

$$U_j(\mathbf{a}, A)\varphi_{j\alpha}(f, t)U_j(\mathbf{a}, A)^{-1} = \sum_{\beta} S_{\alpha\beta}(A^{-1})\varphi_{j\beta}(\{\mathbf{a}, A\}f, t), \tag{4–71}$$

where $A \rightarrow S(A)$ is a matrix representation of the unitary unimodular group SU_2. Suppose the representations U_j possess unique invariant states, Ψ_{j0}:

$$U_j(\mathbf{a}, A)\Psi_{j0} = \Psi_{j0} \qquad j = 1, 2. \tag{4–72}$$

Suppose finally that there exists a unitary operator V such that at time t

$$\varphi_{2\alpha}(f, t) = V\varphi_{1\alpha}(f, t)V^{-1} \tag{4–73}$$

Then

$$U_2(\mathbf{a}, A) = VU_1(\mathbf{a}, A)V^{-1} \tag{4–74}$$

and

$$c\Psi_{20} = V\Psi_{10}, \tag{4–75}$$

where c is a complex number of modulus one.

The proof is a slightly modified version of that of Theorem 3–8. One considers the operator $U_2(\mathbf{a}, A)^{-1}VU_1(\mathbf{a}, A)V^{-1}$ and shows that it is a constant multiple of the identity in $\mathscr{H}_2$. We leave the details to the reader. The relations (4–73) and (4–75) immediately imply the following corollary.

Corollary

In any two theories satisfying the hypotheses of Theorem 4–14 the equal time vacuum expectation values are the same:

$$(\Psi_{10}, \varphi_{1\alpha}(\mathbf{x}_1, t)\ldots\varphi_{1\beta}(\mathbf{x}_n, t)\Psi_0) = (\Psi_{20}, \varphi_{2\alpha}(\mathbf{x}_1, t)\ldots\varphi_{2\beta}(\mathbf{x}_n, t)\Psi_{20}). \tag{4–76}$$

We are now in a position to prove Haag's theorem, which says that if one of the two fields occurring in Theorem 4–14 is a free field so is the other. This will follow easily from a more general result of R. Jost and B. Schroer, so we prove the latter. For simplicity we consider the case of a neutral scalar field, though the result is true more generally.

Theorem 4–15

If $\varphi(x)$ is a hermitian scalar local field for which the vacuum is cyclic, and if

$$(\Psi_0, \varphi(x)\varphi(y)\Psi_0) = \frac{1}{i}\,\Delta^+(x - y;m) \tag{4–77}$$

with $m > 0$ then $\varphi(x)$ is a free field of mass m.

Proof:

If we define $j(x) = (\Box_x + m^2)\varphi(x)$ we see from (4–77) [using $(\Box_x + m^2)\Delta^+(x;m) = 0$] that $(\Psi_0, j(x)j(y)\Psi_0)$ vanishes. Hence $\|j(f)\Psi_0\| = 0$, which leads, by Theorem 4–3, to $j(x) = 0$, i.e., $(\Box + m^2) \times \varphi(x) = 0$. Thus the field φ satisfies the equation of the free field. We must show that it is a free field in the sense of Section 3–2. The main part of the proof consists in showing that the commutator is a multiple of the identity operator.

In momentum space we have $(p^2 - m^2)\tilde{\varphi}(p) = 0$, so the spectrum of $\tilde{\varphi}$ is contained in the hyperboloid $p^2 = m^2$. We can therefore decompose φ into a positive and a negative frequency part in a Lorentz invariant way

$$\varphi(x) = \varphi_+(x) + \varphi_-(x).$$

For example, $\varphi_+(h)$ is defined for a test function h to be

$$\varphi_+(h) = \tilde{\varphi}(\tilde{\theta}\tilde{h}),$$

where $\tilde{\theta}$ is an infinitely differentiable function 0 if $p^2 = m^2$, $p_0 < 0$ and one if $p^2 > 0$, $p_0 > 0$. Since negative energies (positive frequencies!) do not exist

$$\varphi_+(f)\Psi_0 = 0. \tag{4–78}$$

Next consider the state

$$\varphi_+(x)\varphi_-(y)\Psi_0. \tag{4–79}$$

Any momentum occurring in this state is the sum of a forward and a backward time-like vector, each of mass m; so this momentum is space-like or zero. Since there are no states with space-like momentum, (4–79) must be a multiple of Ψ_0. But from (4–78),

$$(\Psi_0, \varphi_+(x)\varphi_-(y)\Psi_0) = (\Psi_0, \varphi(x)\varphi(y)\Psi_0) = \frac{1}{i}\,\Delta^+(x-y).$$

Thus

$$\varphi_+(x)\varphi_-(y)\Psi_0 = \frac{1}{i}\,\Delta^+(x-y)\Psi_0, \tag{4–80}$$

and

$$\frac{1}{i}\,\Delta^+(x-y)\Psi_0 = [\varphi_+(x), \varphi_-(y)]\Psi_0. \tag{4–81}$$

Combining (4–81) with the trivial equation

$$[\varphi_+(x), \varphi_+(y)]\Psi_0 = 0$$

we get

$$[\varphi(x), \varphi(y)]\Psi_0 = \frac{1}{i}\,\Delta(x-y)\Psi_0 + [\varphi_-(x), \varphi_-(y)]\Psi_0, \tag{4–82}$$

where

$$\Delta = \Delta^+ + \overline{\Delta}^+.$$

Let Ψ be any state in the domain of definition of φ, and define

$$F(x, y) = (\Psi, [\varphi_-(x), \varphi_-(y)]\Psi_0).$$

The argument given in the proof of Theorem 4–2 shows that $F(x, y)$ is a tempered distribution. The Fourier transform is 0† unless the variables conjugate to x and y lie in the backward light cone. Therefore, by Theorems 2–6 and 2–7, $F(x,y)$ possess an analytic continuation F into the tube $\operatorname{Im} x$, $\operatorname{Im} y \in V_+$. We can easily see that $F(x,y)$ must vanish if x and y are real, and $(x-y)^2 < 0$. This is obtained from (4–82) by noting that if $(x-y)^2 < 0$, $(1/i)\Delta(x-y) = 0$, and $[\varphi(x), \varphi(y)] = 0$. Thus by analytic continuation (Theorem 2–17) $F(x,y) = 0$ at points of holomorphy and therefore the boundary value F vanishes. Consequently,

$$[\varphi(x), \varphi(y)]\Psi_0 = \frac{1}{i}\,\Delta(x-y)\Psi_0. \tag{4–83}$$

† In this argument, and in several other places in the rest of the proof, we work with the unsmeared fields. This is purely a matter of convenience. The reader can easily supply the required smearing. For example, to deal with (4–79) one considers the tempered distribution T on $\mathbf{R}^8$ defined by

$$(\Psi, \varphi_+(f)\varphi_-(g)\Psi_0) = T(f, g)$$

and argues that its Fourier transform $\tilde{T}(p_1, p_2)$ is zero unless $p_1{}^2 = m^2$, $p_1{}^0 > 0$, $p_2{}^2 = m^2$, $p_2{}^0 < 0$. Thus, in the variable $p_1 + p_2$, its support consists of vectors which are space-like or zero. The conclusion then follows as above.

From (4–83) it follows that the operator $[\varphi(x), \varphi(y)] - (1/i)\Delta(x - y)$, suitably smeared, annihilates the vacuum, and so vanishes, since, again by Theorem 4–3, no annihilation operator can be constructed out of the field operators in a finite region.

In the usual field theory, it is customary to define a free scalar field as one satisfying (4–78) and

$$(\Box + m^2)\varphi(x) = 0, \quad \text{and} \quad [\varphi(x), \varphi(y)] = \frac{1}{i}\Delta(x - y), \tag{4–84}$$

so that in that sense the theorem is already proved. However, it is also easy to verify that the equations (4–84) imply our previous definition of the free field. It suffices to show that the vacuum expectation values are those given by (3–41). From (4–84) we have immediately

$$[\varphi_+(x_1), \varphi(x_j)] = \frac{1}{i}\Delta^+(x_1 - x_j)$$

and therefore

$$\begin{aligned}(\Psi_0, \varphi(x_1)\ldots\varphi(x_n)\Psi_0) &= (\Psi_0, \varphi_+(x_1)\varphi(x_2)\ldots\varphi(x_n)\Psi_0)\\ &= (\Psi_0, [\varphi_+(x_1), \varphi(x_2)]\varphi(x_3)\ldots\varphi(x_n)\Psi_0)\\ &\quad + (\Psi_0, \varphi(x_2)[\varphi_+(x_1), \varphi(x_3)]\varphi(x_4)\ldots\varphi(x_n)\Psi_0)\\ &\quad + \ldots + (\Psi_0, \varphi(x_2)\varphi(x_3)\ldots\varphi(x_{n-1})[\varphi_+(x_1), \varphi(x_n)]\Psi_0),\end{aligned}$$

where we have made use of $\varphi_+(x_1)\Psi_0 = 0$.

Hence

$$\mathscr{W}_n(x_1,\ldots x_n) = \sum_{j=2}^{n} \frac{1}{i}\Delta^+(x_1 - x_j)\mathscr{W}_{n-2}(x_2,\ldots\hat{x}_j,\ldots x_n),$$

where $\hat{x}_j$ means omit x_j. Starting from the two point function we arrive at the formula (3–41) for the vacuum expectation values of the free field. Appealing to the reconstruction theorem we see that $\varphi(x)$ is the free field defined in Section 3–2. ∎

From Theorem 4–15 we get

Theorem 4–16 (*Haag's Theorem*)

Suppose that $\varphi_1(x)$ is a free hermitian scalar field of mass $m > 0$, and $\varphi_2(x)$ is a local field covariant under the inhomogeneous $SL(2,C)$. Suppose further that the fields $\varphi_1(x)$, $\dot{\varphi}_1(x)$, $\varphi_2(x)$, $\dot{\varphi}_2(x)$, satisfy the hypotheses of Theorem 4–14. Then $\varphi_2(x)$ is a free field of mass m.

Proof:

By the corollary to Theorem 4–14 we deduce

$$(\Psi_{20}, \varphi_2(\mathbf{x}, t)\varphi_2(\mathbf{y}, t)\Psi_{20}) = \frac{1}{i}\Delta^+(\mathbf{x}-\mathbf{y}, 0; m). \qquad (4\text{–}85)$$

Any two vectors $(\mathbf{x}, t_1)$ and $(\mathbf{y}, t_2)$ can be brought into the equal time plane $t_1 = t_2$ by a restricted Lorentz transformation if their difference is space-like. Thus combining the given covariance of φ_2 with (4–85) we get by the standard analytic continuation argument

$$(\Psi_{20}, \varphi_2(\mathbf{x}, t_1)\varphi_2(\mathbf{y}, t_2)\Psi_{20}) = \frac{1}{i}\Delta^+(x - y; m).$$

The theorem is then an immediate consequence of Theorem 4–15. ▌

Haag's theorem is very inconvenient; it means that the interaction picture exists only if there is no interaction.

Using the same techniques we can prove a more general result relating any two fields which are unitary equivalent at a given time. It is known as the *generalized Haag's theorem.*

Theorem 4–17

Let there be given two theories as described in the hypotheses of Theorem 4–14, and suppose, in addition, that the theories are invariant under the inhomogeneous $SL(2,C)$, and that some of the operators $\varphi_{j\alpha}$ transform under $SL(2,C)$ according to

$$U_j(a, A)\varphi_{j\alpha}(x)U_j(a, A)^{-1} = \sum S_{\alpha\beta}(A^{-1})\varphi_{j\beta}(Ax + a). \qquad (4\text{–}86)$$

Then the expectation values involving four or fewer of those fields coincide in the two theories.

Remark:

The statement of Theorem 4–17 allows us to apply it to the usual case, where the basic fields are covariant under the inhomogeneous $SL(2,C)$, but their non-covariant conjugate momenta are necessary to form an irreducible set if we restrict ourselves to a given time.

Proof:

Let $W^{(n)}_{j\alpha\ldots\beta}(\xi_1, \ldots \xi_{n-1}), j = 1, 2$ be the holomorphic functions whose boundary values are the left- (respectively right-) hand side of (4–76), where the φ's have been chosen from among the fields satisfying (4–86).

The points at which (4–76) holds are totally space-like and among them are Jost points.† To prove the theorem it suffices to show that these Jost points, together with those which are obtainable from them by real Lorentz transformations, form a real environment for the holomorphic functions $W^{(n)}_{j\alpha\ldots\beta}$ with $n = 2, 3, 4$. (For $n = 1$ the whole affair is trivial.) For $n = 2$ the argument was given after (4–85) in the proof of the previous theorem: any space-like vector ξ can be brought into the form $\xi^0 = 0$ by a real restricted Lorentz transformation. For $n = 3$, any two real space-like vectors ξ_1, ξ_2 can be brought into the equal time plane $\xi_1{}^0 = \xi_2{}^0 = 0$ by a real restricted Lorentz transformation provided the two-plane they span consists of space-like vectors only (if they are colinear we apply the argument given for $n = 2$). The criterion is simply

$$|\xi_1 \cdot \xi_2| < \sqrt{\xi_1{}^2 \xi_2{}^2}, \quad \xi_1{}^2 < 0, \quad \xi_2{}^2 < 0, \tag{4–87}$$

which is surely satisfied by an open set of real Jost points. Finally, for $n = 4$ any three real space-like vectors ξ_1, ξ_2, ξ_3 can be brought into the equal time plane $\xi_1{}^0 = \xi_2{}^0 = \xi_3{}^0 = 0$ by a real restricted Lorentz transformation provided the three-plane they span consists entirely of space-like vectors. Here the criterion is that the matrix

$$\begin{Bmatrix} \xi_1{}^2 & \xi_1 \cdot \xi_2 & \xi_1 \cdot \xi_3 \\ \xi_2 \cdot \xi_1 & \xi_2{}^2 & \xi_2 \cdot \xi_3 \\ \xi_3 \cdot \xi_1 & \xi_3 \cdot \xi_2 & \xi_3{}^2 \end{Bmatrix} \tag{4–88}$$

be negative definite, a condition which is again satisfied by an open set of Jost points. ∎

In Theorem 4–17 we cannot prove the equality of the vacuum expectation values of the two theories for five or more operators by the same method, since four or more vectors cannot be transformed to the equal time plane unless they lie on a special subset, and this is too small to provide a unique analytic continuation for functions defined on it. However the theorem is powerful enough: if two theories are to give different vacuum expectation values for products of 2, 3, or 4 field operators, then inequivalent representations of the canonical commutation relations have to be used.

We believe that the theorems of this section show that the relation between a free field and one with interaction cannot be as simple as the analogy with systems of a finite number of degrees of freedom might

† See Theorem 2–12.

suggest. In particular, the kinematics gets mixed up with the dynamics in the sense that the dynamics determine which representation of the canonical commutation relations we must use. What is even more likely in physically interesting quantum field theories is that equal time commutation relations will make no sense at all; the field might not be an operator unless smeared in time as well as space.

In mentioning these awkward theorems we did not intend to suggest that we think that no sense can be made of equations of motion such as

$$(\Box + m^2)\varphi(x) = \lambda\varphi(x)^3 \tag{4–89}$$

and that their study is not worthwhile. The results of this section merely make clear certain facts of life which must be taken into account in interpreting them. In a certain sense all the theorems of this chapter can be regarded as a girding of the loins preparatory to a study of things like (4–89). To show the equation has solutions we must find a way of defining $\varphi(x)^3$, and to do this we must have some idea of what a quantized field $\varphi(x)$ looks like.

4–6. EQUIVALENCE CLASSES OF LOCAL FIELDS (BORCHERS CLASSES)

One can arrive at the notion of equivalence classes of local fields by several routes. The most direct is the following. Let φ_1 be a hermitian scalar field with cyclic vacuum state. Suppose φ_2 and φ_3 are two other fields with the same representation of the Poincaré group as φ_1. Suppose φ_2 and φ_3 are not necessarily local but instead are *local relative to* φ_1. That means

$$\begin{aligned} [\varphi_1(x), \varphi_2(y)]_- &= 0, \\ [\varphi_1(x), \varphi_3(y)]_- &= 0, \qquad \text{for } (x - y)^2 < 0. \end{aligned} \tag{4–90}$$

Then what can be said about the relation of φ_2 to φ_3? Borchers discovered that they are relatively local:

$$[\varphi_2(x), \varphi_3(y)]_- = 0, \qquad \text{for } (x - y)^2 < 0. \tag{4–91}$$

That immediately implies that they are local (take $\varphi_2 = \varphi_3$). Thus, relatively local fields fall into equivalence classes.†

† We should mention at the outset that what is proved is that if a local φ_1 has the vacuum as cyclic vector, which implies that φ_1 is irreducible, then (4–90) implies (4–91). The φ_2, φ_3 need not be irreducible. Now in the customary mathematical definition of equivalence all elements of an equivalence class must be on the same footing, so strictly speaking it is the *irreducible* relatively local fields which form equivalence classes. In the following, we shall use the term equivalence class somewhat loosely, including reducible relatively local fields in the equivalence class along with the irreducible.

The main purpose of this section is to prove these results and analogous ones involving weak local commutativity. However, before we begin, we want to give some idea of their significance.

What do the fields look like which lie in a given equivalence class? In Section 3–2 we have introduced the notion of a Wick polynomial in a free field. It can be shown that the Wick polynomials in a given free field exhaust the equivalence class of that free field (Ref. 27). It is plausible for fields which are not free that the relations between fields in the same equivalence class will be something like those between the Wick polynomials; they can be expressed as local functions of each other. An example is $\varphi(x)$ and $\varphi(x) + (\Box + m^2)\varphi(x)$. At the moment, the investigation of the precise meaning which should be given to the idea that one field is a local function of another is under intensive investigation but the matter is scarcely in definitive shape. It is closely connected with the theory of von Neumann algebras of open sets mentioned at the end of Section 4–2.

The physical importance of the equivalence relation that arises from relative local commutativity is this: two field theories which have the same Hilbert space $\mathscr{H}$ and transformation law U and whose fields are relatively local have the same S-matrix. The proof of this statement makes use of the Haag–Ruelle theory and so is outside the scope of this book. Nevertheless, because of its importance we want to give some idea of the direction in which the argument goes and at the same time show how equivalence can be practically important. Consider the theory of proton, neutron, and π-mesons, $\pi^{\pm}$ and π^0, already discussed at the end of Section 4–2, in which the proton and neutron fields, ψ_p, ψ_n, together with their adjoints form an irreducible set of operators. In the construction of Haag and Ruelle, one forms states $Q_\alpha(t)\Psi_0$, where $Q_\alpha(t)$ is a suitably smeared polynomial in these fields such that

$$\lim_{t\to\pm\infty} \|Q_\alpha(t)\Psi_0 - \Psi_\alpha^{\substack{\text{out}\\\text{in}}}\| = 0,$$

where the $\Psi_\alpha^{\substack{\text{out}\\\text{in}}}$ are the collision states of neutrons, protons, and π mesons. In this theory, there exists another irreducible set of operators, ψ_n, φ^+, and their adjoints which can create all the states. These fields can also be used to obtain collision states using the Haag–Ruelle prescription. Ruelle's theory assures us that if the fields ψ_p, ψ_n, φ^+, φ^-, φ^0 are local and local relative to one another, then the collision states defined in terms of one irreducible set, ψ_p, ψ_n, are the same as the collision states defined in terms of another ψ_n, φ^+. It follows that the S-matrix is the same whichever set is used for the definition of the asymptotic states.

One gets further perspective on the significance of the equivalence classes of local fields if one considers the following problem: classify field theories with a given $\mathscr{H}$ and U which have different S-matrices. (A really effective solution of this problem would, of course, be a crowning achievement of the general theory of quantized fields even if it did not lead to the calculation of a single cross section.) Suppose we call two field theories *S-equivalent* if they lead to the same S-matrix. Then the preceding can be restated: two field theories in $\mathscr{H}$, with the same U and relatively local fields, are S-equivalent. Thus, S-equivalence classes are made up of Borchers classes. It is not difficult to see that the same S-equivalence class will contain many Borchers classes. For example, from a field theory of a neutral scalar field φ, one can obtain new theory by replacing φ by $V\varphi V^{-1}$, where V is any unitary transformation commuting with the representation U of $\mathscr{P}_+^\uparrow$. In general, $V\varphi V^{-1}$ will not be local relative to φ, but it will yield an S-equivalent theory. It is not yet known whether all the field theories of a given S-equivalence class can be obtained from those of a single Borchers' class by applying all possible V's in this way. What is clear is that a knowledge of the structure of Borchers classes is an essential preliminary in any effort to understand the structure of S-equivalence classes.

With these qualitative remarks completed, we now commence our treatment of equivalence classes. For simplicity, we will discuss field theories given in terms of a single hermitian scalar field. The generalization to arbitrary sets of spinor fields is straightforward. We recall that a field, φ, is weakly local if the equality

$$(\Psi_0, \varphi(x_1)\ldots\varphi(x_n)\Psi_0) = (\Psi_0, \varphi(x_n)\ldots\varphi(x_1)\Psi_0) \tag{4–92}$$

holds for each n and each Jost point $x_1\ldots x_n$. We saw in the *PCT* theorem that φ is weakly local if and only if there exists an antiunitary Θ satisfying

$$\Theta\varphi(f)\Theta^{-1} = \varphi(\hat{f}), \tag{4–93}$$

where

$$\hat{f}(x) = \bar{f}(-x).$$

Since by Theorem 4–5 φ is irreducible, (4–93) determines Θ up to a phase (the proof is in Section 3–5). If there is another irreducible field ψ defined in $\mathscr{H}$ with the same representation U of $\mathscr{P}_+^\uparrow$, it will have a *PCT* operator Θ_ψ, say, if and only if it is weakly local. We then have two *PCT* operators Θ and Θ_ψ; when will they be the same? The reconstruction theorem (Theorem 3–7) and Theorem 4–7 show that if the domains of the two fields satisfy certain conditions, and if φ and ψ *together* satisfy the *WLC* identities, then $\Theta = \Theta_\psi$. We shall describe this situation by saying the fields are *weakly local with respect to each other.* As an example of a field theory with three irreducible fields which are *not* weakly local with respect to each other, consider a non-trivial theory of a hermitian scalar field $\varphi(x)$ which has asymptotic free fields $\varphi^{\text{in}}(x)$, $\varphi^{\text{out}}(x)$, and suppose the theory is asymptotically complete. Since

$$\Theta\varphi^{\text{in}}(f)\Theta^{-1} = \varphi^{\text{out}}(\hat{f}) \neq \varphi^{\text{in}}(\hat{f})$$

the CPT operator Θ for $\varphi(x)$ is not the CPT operator for φ^{in} or φ^{out}. It follows *a fortiori* that $\varphi^{\text{in}}(x)$, $\varphi(x)$, and $\varphi^{\text{out}}(x)$ cannot be local with respect to each other unless all the fields coincide.

We now give a set of conditions that φ and ψ be weakly local relative to each other. These conditions are a subset of those which occur in Theorem 4–7. It should be noted that neither field φ nor ψ of the theorem is assumed local, i.e., neither need satisfy axiom III.

Theorem 4–18

Suppose $\varphi(x)$ is a weakly local field with vacuum as cyclic vector, and suppose $\psi(x)$ is another field transforming under the same representation of $\mathscr{P}_+^\uparrow$, and with the same domain D of definition. Suppose

$$\begin{aligned}(\Psi_0, \varphi(x_1)\ldots\varphi(x_j)\psi(x)\varphi(x_{j+1})\ldots\varphi(x_n)\Psi_0) \\ = (\Psi_0, \varphi(x_n)\ldots\varphi(x_{j+1})\psi(x)\varphi(x_j)\ldots\varphi(x_1)\Psi_0) \quad (4\text{–}94)\end{aligned}$$

holds at Jost points for all j and n; then

(a) $\psi(x)$ is a weakly local field, and

(b) $\varphi(x)$, $\psi(x)$ are weakly local with respect to each other.

Proof:

If Θ is the PCT operator for the field φ, we have for any states Φ, $\Psi \in D$

$$(\Theta\Phi, \Theta\psi(x)\Theta^{-1}\Theta\Psi) = \overline{(\Phi, \psi(x)\Psi)}. \quad (4\text{–}95)$$

Using the invariance of (4–94) under the complex Lorentz group we get for points of holomorphy

$$\begin{aligned}(\Psi_0, \varphi(x_1)\ldots\varphi(x_j)\psi(x)\varphi(x_{j+1})\ldots\varphi(x_n)\Psi_0) \\ -(\Psi_0, \varphi(-x_n)\ldots\varphi(-x_{j+1})\psi(-x)\varphi(-x_j)\ldots\varphi(-x_1)\Psi_0) = 0. \quad (4\text{–}96)\end{aligned}$$

Now both sides of (4–96) can be continued into the tube given by

$$\text{Im}(x_1 - x_2),\ldots\text{Im}(x_j - x), \text{Im}(x - x_{j+1}),\ldots\text{Im}(x_{n-1} - x_n) \in -\mathsf{V}_+ \quad (4\text{–}97)$$

and so the equation holds at all points of holomorphy. Taking the limit from the tube (4–97) we see that (4–96) holds at all real points. If

$$\Psi = \varphi(f_j)\ldots\varphi(f_1)\Psi_0, \qquad \Phi = \varphi(f_{j+1})\ldots\varphi(f_n)\Psi_0 \quad (4\text{–}98)$$

then

$$\Theta\Psi = \varphi(\hat{f}_j)\ldots\varphi(\hat{f}_1)\Psi_0, \qquad \Theta\Phi = \varphi(\hat{f}_{j+1})\ldots\varphi(\hat{f}_n)\Psi_0.$$

Using (4–96) we get

$$(\Psi, \psi(x)\Phi) = (\Theta\Phi, \psi(-x)\Theta\Psi),$$

which when compared with (4–95) leads to

$$\Theta\psi(x)\Theta^{-1} = \psi(-x),$$

an equation valid on states of the form (4–98). Using the hermiticity of ψ, we conclude (just as in the argument at the end of Section 4–1) that it is valid on all of D. Therefore there is a PCT operator for the field $\psi(x)$ [this proves (a)] and this operator is identical with the PCT operator of the field $\varphi(x)$ [this proves (b)]. ▌

We now show that the property of being weakly local with respect to a given irreducible field is *transitive*, in a sense which the theorem states. (The customary definition of transitivity for a relation $\equiv$ is $a \equiv b$ and $b \equiv c$ imply $a \equiv c$. We have this strict transitivity if φ_2 and φ_3 also have the vacuum as cyclic vector.)

Theorem 4–19

Suppose φ_1 is a weakly local field for which the vacuum is cyclic, and φ_j, $j = 2, 3$, are weakly local with respect to φ_1 and have the same domain of definition and representation of $\mathscr{P}_+^\uparrow$. Then φ_2 is weakly local with respect to φ_3.

Proof:

From the proof of Theorem 4–18 it follows that φ_1 and φ_2 have a simultaneous PCT operator, and so have φ_1 and φ_3. Since φ_1 is irreducible this operator is unique; therefore φ_2 and φ_3 have the same PCT operator. Thus they are weakly local with respect to each other, by the PCT theorem. ▌

The important result is that given a weakly local field $\varphi(x)$ for which the vacuum is cyclic, we can form an equivalence class consisting of all fields in the theory which are weakly local with respect to φ. All fields with the same PCT operator and domain are in this class. As far as the S-equivalence of fields in this class is concerned we have the following weak result.

Theorem 4–20

If $\varphi_1(x)$ is weakly local and has Ψ_0 as cyclic vector, and if $\varphi_2(x)$ is weakly local with respect to $\varphi_1(x)$, and if there are asymptotic fields such that

$$\varphi_1^{\text{in}}(x) = \varphi_2^{\text{in}}(x) \tag{4–99}$$

then

$$\varphi_1^{\text{out}}(x) = \varphi_2^{\text{out}}(x). \tag{4–100}$$

Proof:

φ_1 and φ_2 have the same PCT operator, so since Θ carries in-fields into out-fields, (4–100) follows from (4–99). ▮

For local fields there are stronger results, since, as remarked before, (4–99) can be proved using the Haag–Ruelle theory.

The next theorem asserts that the relation of relative locality partitions local fields into classes. This is the result whose significance was discussed at the beginning of this section.

Theorem 4–21

Suppose φ_1, φ_2, and φ_3 are fields with the same domain and representation U of $\mathscr{P}_+^\uparrow$, φ_1 is local and has the vacuum as cyclic vector, and in addition

$$[\varphi_1(x), \varphi_2(y)] = 0 = [\varphi_1(x), \varphi_3(y)] \qquad \text{if } (x - y)^2 < 0$$

then

$$[\varphi_2(x), \varphi_3(y)] = 0 \qquad \text{if } (x - y)^2 < 0.$$

Proof:

The fields φ_1, φ_2, and φ_3 satisfy the conditions of Theorem 4–18, and so φ_2 and φ_3 are weakly local, and weakly local with respect to φ_1. Hence they are weakly local with respect to each other. Thus at Jost points

$$\begin{aligned}(\Psi_0, \varphi_1(x_1)\ldots\varphi_1(x_j)\varphi_2(y)\varphi_3(z)\varphi_1(x_{j+1})\ldots\varphi_1(x_n)\Psi_0)&\\ = (\Psi_0, \varphi_1(x_n)\ldots\varphi_1(x_{j+1})\varphi_3(z)\varphi_2(y)\varphi_1(x_j)\ldots\varphi_1(x_1)\Psi_0)&\\ = (\Psi_0, \varphi_1(x_1)\ldots\varphi_1(x_j)\varphi_3(z)\varphi_2(y)\varphi_1(x_{j+1})\ldots\varphi_1(x_n)\Psi_0)&\end{aligned}$$

since φ_1 is local, and commutes with φ_2 and φ_3. This is precisely the relation studied in the proof of Theorem 4–1, and just as there, we can conclude that for all $x_1\ldots x_n$,

$$(\Psi_0, \varphi_1(x_1)\ldots\varphi_1(x_j)[\varphi_2(y), \varphi_3(z)]\varphi_1(x_{j+1})\ldots\varphi_1(x_n)\Psi_0) = 0 \qquad \text{if } (y - z)^2 < 0.$$

Since Ψ_0 is cyclic for φ_1 we conclude that

$$[\varphi_2(y), \varphi_3(z)] = 0 \quad \text{if } (y - z)^2 < 0, \text{ as required.} \quad \blacksquare$$

Corollary

Putting $\varphi_2 = \varphi_3$ we get, with the same assumptions:
If $\varphi_1(x)$ is local and has Ψ_0 as cyclic vector, and φ_2 is local with respect to φ_1 then φ_2 is local.

Theorem 4–21 shows that relative locality is a transitive property.

As an application of Theorems 4–21 and 4–15 consider the problem of the uniqueness of the solution of the equation

$$(\Box + m^2)u(x) = j(x).$$

Here $j(x)$ is assumed to be a given irreducible local field, and the problem is to find local u which satisfy the equation. Suppose u_1 and u_2 were two such. Since both are local relative to j they are local relative to each other. Thus their difference $\varphi = u_1 - u_2$ is local and satisfies

$$(\Box + m^2)\varphi(x) = 0.$$

φ is therefore a multiple of a free field. We should like to conclude that $\varphi = 0$. To make this step easy we add a further hypothesis. We assume that the free field u_{in} exists, $u_{\text{in}}(f)\Psi_0$ is a one-particle state Ψ_{1f}, and

$$(\Psi_0, u(x)\Psi_{1f}) = (\Psi_0, u_{\text{in}}(x)\Psi_{1f}),$$

for all $f \in \mathscr{S}$. It now follows that

$$(\Psi_0, [u_1(x) - u_2(x)]\Psi_{1f}) = 0,$$

which in turn implies $\varphi = 0$. Thus, we have

Theorem 4–22

If j is a given irreducible local field and $u(x)$ is a local solution of

$$(\Box + m^2)u(x) = j(x) \qquad (4\text{–}101)$$

with a one-particle state at mass m, then u is uniquely determined by the requirement that

$$(\Psi_0, u(x)\Psi_{1f}) = \frac{1}{i}\int \Delta^+(x - y)\, dy\, f(y).$$

There are other hypotheses under which the uniqueness of the local solution of (4–101) follows but we shall not elaborate here.

BIBLIOGRAPHY

The first proof of the global nature of local commutativity is due to R. Jost and O. Steinmann and appeared in

1. A. S. Wightman, "Quantum Field Theory and Analytic Functions of Several Complex Variables," *J. Indian Math. Soc.*, **24**, 625 (1960).

The significance of the polynomial algebra $\mathscr{P}(\mathcal{O})$ associated with a region $\mathcal{O}$ of space-time was first brought out by R. Haag.

2. R. Haag, "Discussion des 'Axiomes' et des propriétés asymptotiques d'une théorie des champs locale avec particules composées" in *Les problèmes mathématiques de la théorie quantique des champs*, CNRS, Paris, 1959.

The fact that the vacuum is cyclic for $\mathscr{P}(\mathcal{O})$ was discovered by Reeh and Schlieder.

3. H. Reeh and S. Schlieder, "Bemerkungen zur Unitäräquivalenz von Lorentzinvarianten Feldern," *Nuovo Cimento*, **22**, 1051 (1961).

The standard theorem on separating vectors for von Neumann algebras is to be found in

4. J. Dixmier, *Les algèbres des opérateurs dans l'espace hilbertien (algèbres de von Neumann)*, Gauthier–Villars, Paris, 1957, p. 6.

The fact that $\mathscr{P}(\mathcal{O})$ does not contain annihilation operators was recognized by a number of authors. See Ref. 16 below for example, or

5. R. Jost, "Properties of Wightman Functions" in *Lectures on Field Theory and the Many-Body Problem*, E. R. Caianiello (ed.), Academic Press, New York, 1961.

The result of Theorem 4–4, that $\{E_0, \mathscr{P}(\mathcal{O})\}$ is irreducible, is contained in Ref. 3. That the irreducibility of the field operators follows from the cyclicity of the vacuum vector was proved in

6. D. Ruelle, "On the Asymptotic Condition in Quantum Field Theory," *Helv. Phys. Acta*, **35**, 147 (1962), Appendix, and also
7. H. J. Borchers, "On the Structure of the Algebra of Field Operators," *Nuovo Cimento*, **24**, 214 (1962). The idea of the proof given here is due to R. Jost.

The *PCT* theorem was first stated as: if a relativistic quantum field theory has space inversion, *P*, as a symmetry, it must have the product of charge conjugation and time inversion, *CT*, as a symmetry also. In this form, it is proved in

8. G. Lüders, "On the Equivalence of Invariance under Time Reversal and under Particle-Anti-Particle Conjugation for Relativistic Field Theories," *Dansk. Mat. Fys. Medd.*, **28**, 5 (1954).

It was Pauli who realized that *PCT* itself is always a symmetry.

9. W. Pauli, "Exclusion Principle, Lorentz Group and Reflection of Space-Time and Charge," p. 30 in *Niels Bohr and the Development of Physics*, W. Pauli (ed.), Pergamon Press, New York, 1955.

The proof given here is that of Jost:

10. R. Jost, "Eine Bemerkung zum CTP Theorem," *Helv. Phys. Acta*, **30**, 409 (1957). This paper was the starting point for many of the applications given in the present chapter. The connection between *PCT* invariance and weak local commutativity introduced here is further discussed in
11. F. J. Dyson, "On the Connection of Weak Local Commutativity and Regularity of Wightman Functions," *Phys. Rev.*, **110**, 579 (1958).

It is interesting that in his paper of 1940, on spin and statistics, Pauli *actually proves what can be regarded as "classical" PCT invariance* for the free field theories he considers: the substitution rule (1–51) is shown to leave the equations invariant if those equations are invariant under the restricted Lorentz group and are linear. The remaining feature of the substitution rule for *PCT* in quantum mechanics, the reversal of order, first appears in the following paper of Schwinger:

12. J. Schwinger, "On the Theory of Quantized Fields I," *Phys. Rev.*, **82**, 914 (1951). However, readers of this paper did not generally recognize that it stated or proved the *PCT* theorem. In the free field case considered by Pauli no operator products occur so the "classical" *PCT* invariance he proved for the equations is the same as the full quantum mechanical *PCT* invariance.

The spin-statistics theorem has a long history. In a fairly general form (general spin, but for free fields) it appears in

13. M. Fierz, "Über die relativische Theorie kräftfreier Teilchen mit beliebigem Spin," *Helv. Phys. Acta*, **12**, 3 (1939).
14. W. Pauli, "On the Connection between Spin and Statistics," *Phys. Rev.*, **58**, 716 (1940).

The first proofs using only the general assumptions of Chapter 3 are due to

15. G. Lüders and B. Zumino, "Connection between Spin and Statistics," *Phys. Rev.*, **110**, 1450 (1958); and
16. N. Burgoyne, "On the Connection of Spin with Statistics," *Nuovo Cimento*, **8**, 807 (1958). Our treatment here follows the latter.

The significant point covered by Theorem 4–8 was first made clear by Dell'Antonio:

17. G. F. Dell'Antonio, "On the Connection of Spin with Statistics," *Ann. Phys.*, **16**, 153 (1961).

The commutation relations between different fields were discussed in the old framework by a number of authors of whom we quote only

18. G. Lüders, "Vertauschungsrelationen zwischen verschiedenen Feldern," *Z. Naturforsch.*, **13a**, 254 (1958). Lüders was the first to use systematically the vector space V described in the text. The notion of Klein transformation which is used by these authors was systematically used by O. Klein in another context:

19. O. Klein, "Quelques remarques sur le traitement approximatif du problème des électrons dans un réseau cristallin par la mécanique quantique," *J. Phys. Radium*, **9**, 1 (1938).

It also appears in

19a. P. Jordan and E. Wigner, "Über das Paulische Äquivalenzverbot," *Z. Physik.*, **47**, 631 (1928).

The treatment of the commutation relations between different fields followed here is due to Araki.

20. H. Araki, "Connection of Spin with Commutation Relations," *J. Math. Phys.*, **2**, 267 (1961).

Haag's theorem originates in

21. R. Haag, "On Quantum Field Theory," *Dan. Mat. Fys. Medd.*, **29**, 12 (1955).

The generalized Haag's theorem was first proved in

22. D. Hall and A. S. Wightman, "A Theorem on Invariant Analytic Functions with Applications to Relativistic Quantum Field Theory," *Mat. Fys. Medd. Dan. Vid. Selsk.*, **31**, 5 (1957).

The simple proof of Theorem 4–15 given here is due to Jost and Schroer (Ref. 5). A different treatment is found in

23. O. W. Greenberg, "Haag's Theorem and Clothed Operators," *Phys. Rev.*, **115**, 706 (1959).

A proof of Haag's theorem similar to that of Jost and Schroer was given by Federbush and Johnson:

24. P. G. Federbush and K. A. Johnson, "The Uniqueness of the Two-Point Function," *Phys. Rev.*, **120**, 1926 (1960).

Equivalence classes of local fields were first described in

25. H. J. Borchers, "Über die Mannigfaltigkeit der interpolierenden Felder zu einer kausalen *S*-Matrix," *Nuovo Cimento*, **15**, 784 (1960).

Examples of relatively local fields which have the same *S*-matrix when evaluated in perturbation theory were found by Chisholm, Salam, and Kamefuchi. References to their papers can be found in

26. S. Kamefuchi, L. O'Raifeartaigh, and Abdus Salam, "Change of Variables and Equivalence Theorems in Quantum Field Theories, *Nucl. Phys.*, **28**, 529 (1961).

The equivalence class of a free hermitian scalar field was determined independently by B. Schroer (unpublished) and H. Epstein.

27. H. Epstein, "On the Borchers Class of a Free Field," *Nuovo Cimento*, **27**, 886 (1963).

The literature on statistics different from Fermi or Bose statistics can be traced from the references in

28. O. W. Greenberg and A. Messiah, "Are There Particles in Nature Other Than Bosons or Fermions?" *Phys. Rev.*, **136**, B248 (1964).

For a systematic account of the Haag-Ruelle theory, and for further study, see

29. R. Jost, *General Theory of Quantized Fields* (*Lectures in Applied Mathematics IV*: Proceedings of the Summer Seminar, Boulder, Colorado, 1960), American Math. Soc., Providence, R.I., 1965.

APPENDIX

SOME MORE RECENT DEVELOPMENTS IN QUANTUM FIELD THEORY

In the decade and a half since this book was originally written, there has been important progress in the foundations of quantum field theory. A systematic review would be out of place here; it would require an addition whose length would be a substantial fraction of the book. Furthermore, the interested reader has available the systematic review of Streater [1] as well as the treatise of Bogolubov, Logunov, and Todorov [2], both of which have extensive bibliographies. The purpose of this Appendix is rather to indicate three developments which have changed our outlook on the foundations of field theory as described in the text. These are constructive quantum field theory, the theory of local algebras, and the theory of superselection rules.

CONSTRUCTIVE QUANTUM FIELD THEORY AND THE EXISTENCE OF NON-TRIVIAL THEORIES OF INTERACTING FIELDS

In the introduction we described the Main Problem of quantum field theory as: "either to show that the idealizations involved in the fundamental notions of the theory (relativistic invariance, quantum mechanics, local fields, etc.) are incompatible in some physical sense, or to recast the theory in such a form that it provides a practical language for the description of elementary particle dynamics." It was known in 1964 that the axioms given in Chapter 3 are consistent because they are satisfied by theories of free fields. On the other hand, at that time not a single example of a theory of interacting fields satisfying the axioms was known. Thus a natural first step toward a solution of the Main Problem was the construction of examples of theories of interacting fields satisfying the axioms.

Lagrangian quantum field theory offers an abundant supply of models whose solutions *ought* to satisfy the axioms. Constructive quantum field theory deals with the mathematical constructions necessary to prove that such solutions exist. It had its beginnings in the early 1960s and by the middle 1970s had become a well-developed industry.

It was preceded by the pioneering studies of Friedrichs [3] and Segal [4] on the mathematical foundations of quantum field theory, but the work of these authors did not treat models with coupled quantized fields. Friedrichs' later book on perturbation theory [5] was an inspiration for the work done on models in the late 1960s. (The possibilities and some of the difficulties of such a program were reviewed in [6] at the outset of constructive quantum field theory. The survey [7] discusses the general significance of axiomatic and constructive field theory. Part of the following recapitulates some of the material in these earlier articles; the reader will find more detail and a more complete bibliography there.)

The first results obtained in constructive quantum field theory were proofs of the existence of solutions for simplified versions of some Lagrangian models in which cutoffs are introduced. For example, the first model considered was a cutoff version of the model, Y_4, in which a spinor field, ψ, and a scalar field, φ, interact via a Yukawa interaction given by

$$H_{\mathrm{Y}} = \lambda \int :\psi^{+}(x)\psi(x): \varphi(x)\, d^3x,$$

the total Hamiltonian being given by

$$H = H_{0\mathrm{F}} + H_{0\mathrm{B}} + H_{\mathrm{Y}}.$$

Here $H_{0\mathrm{F}}$ and $H_{0\mathrm{B}}$ are the Hamiltonians for a free spinor field of mass M_0 and for a free boson field of mass m_0, respectively [8, 9] and $\psi^{+} = \psi^{*T}\gamma_0$. A cutoff version of this model is obtained by expanding the fields in Fourier series in a cubic box, V, of volume $|V|$ and keeping only Fourier components of frequencies $< K$. The corresponding interaction is

$$H_{\mathrm{Y;V,K}} = \lambda \int_V :\psi_{\mathrm{K}}^{+}(x)\psi_{\mathrm{K}}(x): \varphi_{\mathrm{K}}(x)\, d^3x.$$

It is customary to replace $H_{0\mathrm{F}}$ and $H_{0\mathrm{B}}$ by $H_{0\mathrm{F,V}}$ and $H_{0\mathrm{B,V}}$, respectively, where the latter are the Hamiltonians for free particles confined to the box V with periodic boundary conditions. The cutoff total Hamiltonian

$$H_{\mathrm{V,K}} = H_{0\mathrm{F,V}} + H_{0\mathrm{B,V}} + H_{\mathrm{Y;V,K}}$$

then describes a system of an infinite number of degrees of freedom of which only a finite subset is coupled through the interaction $H_{\mathrm{Y;V,K}}$. For such a Hamiltonian, the interaction is small compared with the free Hamiltonian in the sense of T. Kato, and a standard theorem asserts that the total Hamiltonian is self-adjoint on the domain of the free Hamiltonian [10]. The self-adjointness $H_{V,K}$ guarantees that

$$U_{V,K}(t) = \exp(itH_{V,K})$$

defines a continuous unitary one-parameter group. $U_{V,K}(t)$ describes the time development of states and operators in the cutoff theory. Another important result of the study of this cutoff model is the existence of vacuum expectation values of products of field operators:

$$\left(\Psi_{0,V,K}, \prod_{j=1}^{n} \psi_K(t_j, \mathbf{x}_j) \prod_{k=1}^{n} \psi_K^+ (t_k, \mathbf{x}_k) \prod_{l=1}^{m} \varphi_K(t_l, \mathbf{x}_l)\, \Psi_{0,V,K}\right). \quad (A.1)$$

Here $\Psi_{0,V,K}$ is the ground state of $H_{V,K}$. It is non-degenerate for sufficiently small λ. For general λ it cannot be excluded that the ground state has a non-trivial degeneracy. If that occurs, one can define more than one quantity (A.1).

The simplicity of this cutoff version of Y_4 arises in part from the fact that the interaction is only linear in the boson field, φ_K, while the free Hamiltonian H_{0B} is quadratic. Models of a self-coupled boson field φ_K, in which the coupling is of higher degree in φ_K than the free Hamiltonian, were studied in [11]. Here the Hamiltonian is again bounded below but only because one insists that the interaction

$$H_{I,V,K} = \lambda \int_V \mathscr{P}(\varphi_K(x))\, d^3\mathbf{x}$$

contain a polynomial $\mathscr{P}$ which is bounded below, and $\lambda \geqslant 0$. The theory, whose Hamiltonian is

$$H_{0B} + H_{I,V,K},$$

describes an infinite number of oscillators, a finite number of which are coupled. It may be referred to for brevity as cutoff $\mathscr{P}(\varphi)_4$ theory. For it, one can establish the non-degeneracy of the ground state for all coupling strengths, $\lambda \geqslant 0$, and the existence of a set of vacuum expectation values

$$\left(\Psi_{0,V,K}, \prod_{j=1}^{n} \varphi_K(t_j, \mathbf{x}_j)\, \Psi_{0,V,K}\right) \quad (A.2)$$

These results for cutoff Y_4 and cutoff $\mathscr{P}(\varphi)_4$ theories were generalized to a wide class of cutoff theories of coupled fermion and boson fields in [12].

The results just described leave entirely open the question of the existence of a limiting theory as $|V| \to \infty$ and $K \to \infty$. However, evidence from statistical mechanics and from the study of the perturbation series makes the following two conclusions very plausible.

(a) The limit as $K \to \infty$ can only exist in general if extra terms, which become infinite as $K \to \infty$, are added to the Hamiltonian. The addition of these terms can often be interpreted in terms of a renormalization of the basic parameters of the theory.

(b) Not all quantities associated with the theory may be expected to have well-defined limits as $|V| \to \infty$ and $K \to \infty$. However, the vacuum expectation values (A.1) and (A.2) are reasonable candidates for convergence, since each term of the perturbation series for them is known to converge as $|V| \to \infty$ and $K \to \infty$, once the renormalizations mentioned under (a) have been carried out.

The second logical step toward the construction of solutions was then to prove that such limits exist. However, what actually happened was more complicated. The process of renormalization was first controlled in an even simpler model, in which the $|V| \to \infty$ limit posed no problem at all. This was a theory of non-relativistic fermions coupled to relativistic mesons. Here the Hamiltonian is

$$H_K = (2M)^{-1} \int \nabla\psi^*(\mathbf{x}) \cdot \nabla\psi(\mathbf{x})\, d^3\mathbf{x} + H_{0,B} + g \int \psi^*(\mathbf{x})\psi(\mathbf{x})\varphi_K(\mathbf{x})\, d^3x.$$

The fermion field ψ describes non-relativistic nucleons whose number

$$N = \int \psi^*(\mathbf{x})\psi(\mathbf{x})\, d^3\mathbf{x}$$

is exactly conserved

$$[H_K, N] = 0.$$

The renormalization process corresponds to the replacement of H_K by $H_\infty = \lim_{K\to\infty}(H_K - N\,\Delta E_K)$ where ΔE_K is the lowest-order self-energy of a nucleon arising from the meson–nucleon interaction. It was shown that $\exp[it(H_K - N\,\Delta E_K)]$ converges to the continuous unitary one-parameter group $\exp(itH_\infty)$ [13]. (This is more than one expects for general models.) Furthermore, the vacuum expectation values (A.1) converge [14].

The preceding results on the non-relativistic model represented a fundamental step forward because they give a non-perturbative treatment of a non-trivial renormalization. However, they did not involve the full complications of a relativistic theory because the model does not predict a non-trivial vacuum polarization. The next stage in the development of the theory had to deal with the phenomenon of vacuum polarization and required essentially new ideas. The technical difficulties were so great that several workers had the genial idea to consider the models first in space-times of lower dimension, where the singularities of the fundamental solutions of the wave equation are less violent [15–19].

The $\lambda(\varphi^4)_2$ model has, if one believes the evidence of its solution by perturbation series in λ, no ultraviolet divergences, but does have non-trivial vacuum polarization. Thus, it is a suitable test case for existence theorems. The Hamiltonian of the $\lambda(\varphi^4)_2$ model in a box is a well-defined hermitian operator when the φ^4 is understood as Wick-ordered. However, the Hamiltonian is not obviously bounded below, since the Wick-ordering destroys the formal positivity of φ^4.

This problem was solved in a paper [15] which contained a number of mathematical ideas which were important for later developments. In particular, it first used the probabilistic methods later associated with Euclidean field theory to study the semi-group $\{\exp(-tH_{K,V});\ 0 \leqslant t < \infty\}$. In the realization of H_V thereby obtained, $H_{I,V} = \lambda \int_V :\varphi^4: dx$ is a multiplication operator on a space of square-integrable functions. It is not bounded below, but the sets on which it is large and negative are so small in measure that the addition of the positive H_0 makes the sum $H_V = H_0 + H_{I,V}$ bounded below.

The next most difficult case after the $\lambda(\varphi^4)_2$ model is Y_2, the theory of a scalar or pseudo-scalar boson field coupled to a fermion field. It requires an infinite boson mass renormalization and an infinite vacuum energy renormalization. The Hamiltonian of Y_2 was constructed and proved bounded below in [16, 17, 19].

For $\lambda(\varphi^4)_3$, the φ^4 theory in three dimensional space-time, the situation is even more complicated. The ultraviolet divergences are sufficiently bad that the Fock representation of the commutation relations cannot be used to make the Hamiltonian in a box well defined: a unitarily inequivalent representation has to be used, constructed by a process referred to as *wave function renormalization* [20]. (For a discussion of a class of models which puts wave function renormalization in a more general context, see [21].) The proof that the renormalized $\lambda(\varphi^4)_3$ Hamiltonian is bounded below remained an open problem for several years until finally, by an extraordinary effort, it was solved in [22].

The work quoted in [15–22] shows that a renormalized Hamiltonian in a box but free of ultraviolet cutoffs and bounded below can be constructed for the $\lambda(\varphi^4)_2$, Y_2, and $\lambda(\varphi^4)_3$ models. The next problem was to get rid of the box. Here the decisive ideas were these [23, 18, 24]:

(c) Because of Haag's theorem, one cannot expect an interaction term such as $\lambda \int :\varphi^4: dx$ integrated over all space to make sense, even for the least singular case of two dimensions. However, one does expect

$$H_{\rm I}(g) = \lambda \int dx\, g(x) :\varphi^4: (x)$$

to make sense if $g(x)$ is a smooth positive function vanishing outside some bounded region. If $g(x) = 1$ in some region, then

$$H(g) = H_0 + H_{\rm I}(g)$$

ought to serve as a correct *local* Hamiltonian in the dependence domain of the region where $g = 1$ (see Figure A.1). That means that

$$\varphi(t, \mathbf{x}) = \exp(iH(g)t)\varphi(0, \mathbf{x}) \exp(-itH(g))$$

should be independent of g and give the correct time dependence of the field φ as long as $\{t, \mathbf{x}\}$ stays in the diamond indicated in Figure A.1.

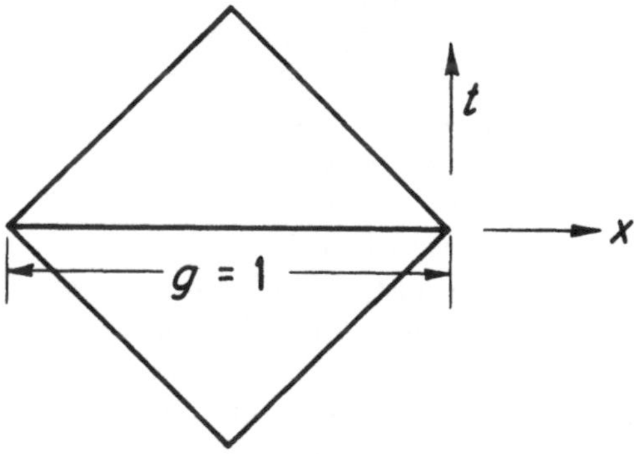

Figure A.1. The dependence domain of the region where $g = 1$.

(d) Although $H(g)$ does not converge as $g \to 1$, the family of such $H(g)$ defines a unique automorphism of local algebras. That means that if F is a bounded function of the smeared fields

$$\varphi(f) = \int dt\, dx\, f(t, x)\varphi(t, x)$$

with f having support in some bounded region, $\mathcal{O}$, then

$$\alpha_\tau F(\varphi(f)) = F(\varphi(f_\tau)),$$
$$f_\tau(\tau, x) = f_\tau(t, x)$$

is independent of g as soon as the region where g is 1 gets sufficiently large in both directions. The result is then to be interpreted: one has the *right local algebras* and the *right automorphism of them* but the representation is the wrong representation because the automorphism α_τ is not an inner automorphism; i.e., there is no unitary operator $V(\tau)$ such that $\alpha_\tau(A) = V(\tau)AV(\tau)^*$. This implies, in particular, that there is no self-adjoint operator H such that

$$\alpha_\tau(A) = \exp(iH\tau)A\exp(-iH\tau).$$

(e) To get the right representation one must construct a vacuum functional on the local algebras, and then appeal to the Gelfand–Naimark–Segal (GNS) theorem. This asserts that, to every functional, ρ, there is a vector Ψ_ρ in some Hilbert space $\mathscr{H}_\rho$, such that ρ is given by the expectation value in the state Ψ_ρ. The theorem, which is stated precisely on p. 192, actually constructs $\mathscr{H}_\rho$ and the representation of the algebra in question, in a manner similar to the "Reconstruction Theorem" of Section 3.4; see [67]. The idea is to use the local Hamiltonians to construct approximate vacuum states. The state, $\Psi_{0,V}$, is now chosen as the ground state, $\Psi_0(g)$, of the $H(g)$; it is proved non-degenerate. One regards $(\Psi_0(g), A\Psi_0(g)) = \rho_g(A)$ as a linear functional on the local algebras and looks for convergent sequences ρ_{g_i}. The limiting states and their associated representations are candidates for the correct physical representation in which the axioms are satisfied.

The extraordinary theoretical structure which was erected to fulfill these ideas is reviewed in [25] and [26]. It is sometimes referred to as the *Hamiltonian strategy* in constructive quantum field theory. The results obtained for $(\varphi^4)_2$, the φ^4 theory in two-dimensional space-time, were soon extended to $\mathscr{P}(\varphi)_2$ where $\mathscr{P}$ is any polynomial bounded below [18, 27].

We have already noted that in the work just quoted, it was found useful to study a Hamiltonian such as $H(g)$ via the associated semigroup $\{\exp[-t(H(g))]; 0 \leqslant t < \infty\}$. In the context of a relativistic theory this can be regarded as a Euclidean method since the substitution $t \to it$ converts the Minkowski metric $c^2t^2 - \mathrm{x}^2$ into a Euclidean metric. Now, parallel to the Hamiltonian strategy, there had developed in the 1960s the *Euclidean strategy* in which one works with imaginary time vacuum expectation values (or Green's functions). The idea, fundamental for the Euclidean strategy, that the analytic continuation of the Green functions to imaginary times should define the correlation functions of a Euclidean field theory goes back to Schwinger and Nakano [28, 29]. It was extensively explored by Symanzik [30–32].

The beauty of the Euclidean formalism is that it promises to give mathematical meaning to the celebrated solution of quantum field theory by quadratures. The formula in question is, for the case of a single self-coupled Bose field,

$$\langle\varphi(x_1)\cdots\varphi(x_k)\rangle = Z^{-1}\int\prod_{j=1}^{k}\varphi_{\mathrm{E}}(x_j)\,d\mu_{m_0^2}(\varphi_{\mathrm{E}})\exp\left[\int\mathscr{L}_{\mathrm{I}}(\varphi_{\mathrm{E}}(x))\,d^nx\right]$$

(commonly known as the Euclidean Gell-Mann–Low formula). Here n is the space-time dimension

$$Z = \int d\mu_{m_0^2}(\varphi_{\mathrm{E}})\exp\left[\int\mathscr{L}_{\mathrm{I}}(\varphi_{\mathrm{E}}(x))\,d^nx\right]$$

and $d\mu_{m_0^2}(\varphi_{\mathrm{E}})$ is the Gaussian probability measure of mean zero and covariance

$$\int\varphi_{\mathrm{E}}(x)\varphi_{\mathrm{E}}(y)\,d\mu_{m_0^2}(\varphi_{\mathrm{E}}) = ([-\Delta+m_0^2]^{-1})(x,y).$$

(For an introduction to the required notions of probability theory see [33].) $\mathscr{L}_{\mathrm{I}}(\varphi_{\mathrm{E}})$ is the interaction Lagrangian density of the theory; for the $\lambda\mathscr{P}(\varphi)_2$ theory, for example, it is $-\lambda\mathscr{P}(\varphi_{\mathrm{E}}(x))$.

The formula for the Schwinger functions does not make sense as it stands, but if an ultraviolet and a box cutoff are introduced, it is perfectly well defined and can serve as a starting point for the construction of the theory free of cutoffs.

What was lacking in Euclidean field theory, even at the end of the 1960s, was a reconstruction theorem which would permit one to pass from the solution of a Euclidean field theory to a corresponding quantum field theory in Minkowski space, satisfying assumptions 0, I, II, and III of Chapter 3 (pp. 96–102).

Such a reconstruction theorem was provided in [34], where it was shown that there is a natural extension to Euclidean field theory of the Markov property usually defined for stochastic processes, and that it, together with a few other properties, suffices for the return via analytic continuation from the Euclidean correlation functions (Schwinger functions) to a set of vacuum expectation values for a theory satisfying 0, I, II, and III.

Two other papers also seem to have provided a considerable encouragement to people to think Euclidean. In [35] it was shown that the existence of a vacuum energy per unit volume in $\mathscr{P}(\varphi)_2$ theories follows from the striking symmetry (Nelson symmetry)

$$(\Omega_0, \exp(-tH_l)\Omega_0) = (\Omega_0, \exp(-lH_t)\Omega_0)$$

where

$$H_l = H_0 + \int_{-l/2}^{l/2} dx :\mathscr{P}(\varphi(x)):$$

and Ω_0 is the ground state of H_0. This identity is an immediate consequence of the fact that both expressions in question can be written as functional integrals in which the dependence on the integration variable, φ_{E}, appears in the form of a Euclidean invariant integral of the function

$$\exp\left[-\int_{-t/2}^{t/2}\int_{-l/2}^{l/2} d^2x :\mathscr{P}(\varphi_{\mathrm{E}}(x)):\right].$$

This is a striking example of an identity which becomes obvious when it is expressed in appropriate Euclidean language. The second paper was [36], which gave conditions on Schwinger functions guaranteeing that they should give rise to a quantum field theory in Minkowski space. Since these conditions are eminently practical, they gave a strong incentive to the study of Euclidean field theory models.

In Euclidean field theory the relation between statistical mechanics and field theory is more than an analogy. For boson fields the Schwinger functions *are* correlation functions of a certain (somewhat peculiar!) statistical mechanics. This suggests that the methods by which the thermodynamic limit is controlled in statistical mechanics should be adaptable to Euclidean field theory.

A striking application of these ideas is obtained if one approximates the Euclidean φ^4 theory by a lattice theory in which Euclidean space is replaced by a lattice of points. The free Hamiltonian then is chosen as (the discrete analogue of) the Laplacian with Dirichlet boundary conditions, which turns out to be effectively a ferromagnet. For $\mathscr{P}$ an even polynomial plus a linear term, it can then be shown that the Schwinger functions converge monotonically as the lattice is increased in size, independently of the lattice spacing [37, 38].

Another result which is plausible in view of the analogy with the statistical mechanics of ferromagnets is that of [39] in which it is shown that for an interaction $\lambda\mathscr{P}(\varphi) + \mu\varphi^k$ with $k < \deg\mathscr{P}$, where k is odd there is a unique vacuum and mass gap for all sufficiently large $|\mu|$. This is analogous to the behavior of a ferromagnet: when a ferromagnet is put in a magnetic field, its magnetization and hence its state is uniquely determined.

For some of the closer comparisons between ferromagnets and $\mathscr{P}(\varphi)_2$ models and, in particular, for the relation to the Ising model, it is necessary to restrict the polynomial to one of fourth degree, which without loss of generality can be taken as $\mathscr{P}(\varphi) = \lambda\varphi^4 + \mu\varphi^2 + \nu\varphi$. In [40, 41] it is shown that there is a unique vacuum for $\nu \neq 0$.

One of the most beautiful results along this line is the proof that for $\nu = 0$ and all sufficiently large negative μ the $\mathscr{P}(\varphi) = \lambda\varphi^4 + \mu\varphi^2$ theory has at least two distinct solutions, one in which the expectation value of φ in the ground state, $\langle\varphi(x)\rangle$, is >0 and another for which it is <0 [42]. Combined with the preceding result, this suggests that the phase diagram of the $\mathscr{P}(\varphi) = \lambda\varphi^4 + \mu\varphi^2 + \nu\varphi$ theory in the λ–μ plane is similar to that of the Ising model ferromagnet in the β–B plane where $\beta = (kT)^{-1}$, k being Boltzmann's constant, T the absolute temperature, and B the external magnetic field.

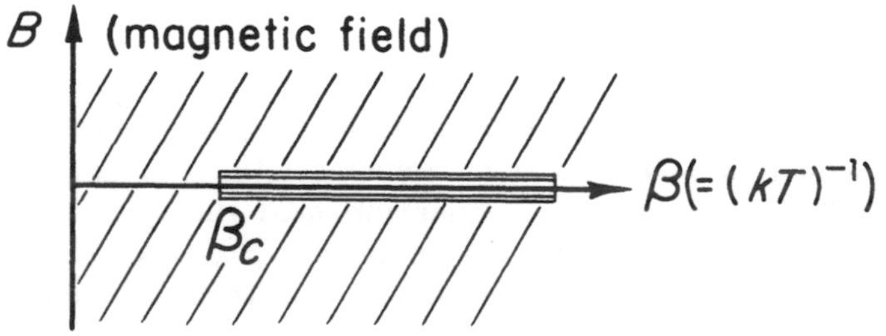

Figure A.2. Phase diagram of the Ising ferromagnet (nearest-neighbor interaction). The phase is unique and the correlation functions decay exponentially for all values of β and B except those satisfying $B = 0$, $\beta_c \leqslant \beta < \infty$, where there are exactly two phases. The phase diagram for the $\lambda\varphi^4 + \mu\varphi^2 + \nu\varphi$ model is similar, with B replaced by ν and β by λ. The solution is known to be unique for $\nu \neq 0$ and for all sufficiently small positive λ. It is known that at least two solutions, I and II, exist for all sufficiently large λ with $\varphi \to -\varphi$ symmetry broken in the sense that $(\Psi_{0\mathrm{I}}, \varphi_{\mathrm{I}}(x)\,\Psi_{0\mathrm{I}}) = -(\Psi_{0\mathrm{II}}, \varphi_{\mathrm{II}}(x)\,\Psi_{0\mathrm{II}}) \neq 0$.

It is time to cease celebrating the successes of Euclidean field theory in relating the ideas of statistical mechanics to quantum field theory and return to the Main Problem. The books [43] and [44] give a more complete survey, with references complete up to their dates of publication.

From the point of view of axiomatic quantum field theory, one of the most important results achieved (by a combination of the Hamiltonian and Euclidean strategies) was the verification for the $\lambda\mathscr{P}(\varphi)_2$ model of assumptions 0, I, II, III of Chapter 3, i.e., the proof that *the* $\lambda\mathscr{P}(\varphi)_2$ *model defines (on two-dimensional space-time) a local relativistic quantum field theory with a cyclic vacuum state.* This was established in [45] for any $\mathscr{P}$ bounded below and all sufficiently small λ. For $\mathscr{P}$ even plus a linear term but for all λ, it was established in [37, 38, 44]. An alternative treatment of this case, based on a very neat and powerful formalism expressing the physical content of the theory in terms of a generating functional for the Schwinger functions, was given in [46, 47]. This justified the faith of those who had, for so many years, believed in the existence of non-trivial local relativistic quantum field theories.

Of course, the skeptics could still have questions:

Could the constructed solution be, in fact, trivial?

Does the constructed solution have anything to do with the perturbation series for the vacuum expectation values traditionally associated with the $\mathscr{P}(\varphi)_2$ model?

Could asymptotic completeness (assumption IV, p. 102) fail for the model?

Might the existence of a solution be a freak of two dimensions or of the regularity of the interaction; could the analogous problems for higher dimensions or for more singular interactions be without non-trivial solutions?

An important step toward answering these questions was already taken at the same time that axioms 0, I, II, and III were verified for small positive λ [45]. It was shown that the mass spectrum of the theory has the form generally expected: Above the unique vacuum state there is a gap, then the single-particle states of mass m, and then another gap. Further, the field has non-vanishing matrix elements, $(\Psi_0, \varphi(x)\Psi_1)$, between the vacuum, Ψ_0, and one-particle states, Ψ_1. This information provides the necessary input for the Haag–Ruelle theory of scattering [48, 49]. The Haag–Ruelle theory constructs collision states corresponding to ingoing and outgoing beams with arbitrary numbers of the particles of mass m whose single-particle states occurred in the hypotheses. The scattering amplitudes are given in terms of the Fourier transforms of Green's functions evaluated on the mass shell. It was shown in [50] that when smeared with appropriate test functions, the Schwinger functions for $\lambda\mathscr{P}(\varphi)_2$ are infinitely differentiable for $0 \leqslant \lambda < \lambda_0$ with λ_0 some strictly positive number, and that their Taylor series are given by the usual Feynman rules. Later on it was shown that for $\deg \mathscr{P} = 4$ the series is Borel summable to the exact solution [51]. An additional argument is necessary for the real time Green functions and the S-matrix elements themselves. This is given in [52–54]. These results completely answer the first two questions: *The solution of the $\lambda\mathscr{P}(\varphi)_2$ model is non-trivial and its scattering matrix and Green's functions have perturbation series in λ that agree with those given by the standard Feynman rules.*

The answer to the third question is very likely no: The $\lambda\mathscr{P}(\varphi)_2$ theory is almost certainly *not* asymptotically complete, at least for large λ. To explain the reasons for this statement (which is shocking if one guides oneself entirely by the results of formal perturbation theory) is the principal task of the last part of this Appendix. First, we will continue with developments bearing on the last question, what constructive field theory has to say about higher dimensions and other interactions.

We have already remarked that the renormalized Y_2 Hamiltonian in a box was early shown to be bounded below. The Hamiltonian strategy required a sharper result: $H(g)$ for Y_2 is self-adjoint [55]. The further development of the Y_2 model theory followed the Hamiltonian strategy and culminated in the construction of a theory (or theories) free of cutoffs [25, 26]. It was then natural to ask how the Y_2 theory could be developed using the Euclidean strategy. A crucial first step was taken in [57], where it was shown that there is a meaningful Euclidean Gell-Mann–Low formula for the Schwinger functions of Y_2 with the fermions integrated out

$$\left\langle \prod_{j=1}^{r} \psi(x_j) \prod_{k=1}^{r} \psi^+(y_k) \prod_{l=1}^{s} \varphi(z_l) \right\rangle$$

$$= Z^{-1} \int d\mu_{m_0{}^2}(\varphi_{\mathrm{E}}) \, [\det_{j,k} S_{\mathrm{E}}(x_j, y_k; g\varphi_{\mathrm{E}})] \left[\prod_{l=1}^{s} \varphi_{\mathrm{E}}(z_l) \right] \det_{\mathrm{ren}}[1 + S_{\mathrm{E}} * g_{\mathrm{E}}]$$

with

$$Z = \int d\mu_{m_0{}^2}(\varphi_{\mathrm{E}}) \det_{\mathrm{ren}}[1 + S_{\mathrm{E}} * g\varphi_{\mathrm{E}}].$$

Here $\det_{\mathrm{ren}}(1 + S_{\mathrm{E}} * g\varphi_{\mathrm{E}})$ is a suitably defined renormalized Fredholm determinant of the operator $1 + S_{\mathrm{E}} * g\varphi_{\mathrm{E}}$ where $S_{\mathrm{E}} * g\varphi_{\mathrm{E}}$ is the integral operator whose kernel is

$$S_{\mathrm{E}}(x-y)\, g(y)\, \varphi_{\mathrm{E}}(y)$$

$S_{\mathrm{E}}(x)$ being the fundamental solution of the Euclidean Dirac equation. g is a smooth function, 1 in a large region and 0 outside a somewhat larger region, which describes the box in which interaction takes place. More precisely, it was shown in [57] that the integrands of the two indicated integrals are in $L^p(d\mu_{m_0{}^2})$ for all $1 \leqslant p < \infty$. Many of the previously obtained results for Y_2 as well as new results have been extracted from this formula. For example, the axioms 0, I, II, and III have been verified for sufficiently weak coupling [58, 59]. For further results and references see [60].

The situation for $\lambda(\varphi^4)_3$ is somewhat similar. Axioms 0, I, II, and III have been verified for all sufficiently weak coupling and the Euclidean

Gell-Mann–Low formula has been made to work [61, 62, 60]. These results show that theories exist satisfying the axioms and having infinite renormalizations, and furthermore, that non-trivial theories can exist in space-times of at least three dimensions.

Unfortunately, none of the papers gives a complete answer to the question of the existence of solutions for typical four-dimensional theories, say, $(\varphi^4)_4$ or the quantum electrodynamics of spin $\frac{1}{2}$ particles. The phenomenon of charge renormalization, which is unavoidable for the treatment of such renormalizable but not superrenormalizable theories, has as yet not been treated in constructive field theory. It is clearly one of the main roadblocks in the way of a completely satisfactory answer to the question of the existence of non-trivial theories. There are many intriguing but so far inconclusive investigations in this area, of which we mention only [63, 64].

Thus, in our opinion, *constructive quantum field theory has provided a very satisfactory solution of the Main Problem; but, so far, only for superrenormalizable theories in space-times of dimension* $\leqslant 3$.

LOCAL ALGEBRAS AND SUPERSELECTION SECTORS

In the early 1960s, an alternative version of the foundations of relativistic quantum field theory was developed [65], based on the use of algebras of bounded operators. Although in principle this formalism should be roughly equivalent to the axiomatic field theory described in this book, it turned out to be very difficult to obtain a smooth connection between the two, and so it has been developed in parallel with the quantum theory of fields [66]. For certain general theoretical purposes, the local algebra formalism turned out to be very natural. In particular, it yields a theory of superselection rules which throws new light on the axioms discussed in this book. In the following we sketch this special line of argument and refer the reader to [67] for a systematic account of the algebraic approach and a bibliography.

The basic objects of the local algebra formalism are C^* algebras, $\mathscr{A}(\mathscr{O})$, associated with bounded open sets, $\mathscr{O}$, in space-time. These algebras are supposed to form a net in the sense that

I. $\mathscr{O}_1 \subseteq \mathscr{O}_2$ implies $\mathscr{A}(\mathscr{O}_1) \subseteq \mathscr{A}(\mathscr{O}_2)$.

The relativistic invariance of the theory is expressed as

II. There exists a representation, α, of the Poincaré group by automorphisms of $\mathscr{A}$: $\{a, \Lambda\} \to \alpha(a, \Lambda)$ such that if $A \in \mathscr{A}(\mathscr{O})$, then $\alpha(a, \Lambda)(A) \in \mathscr{A}(\Lambda\mathscr{O} + a)$.

It therefore makes sense to speak of $\bigcup_{\mathcal{O}} \mathscr{A}(\mathcal{O})$ and its norm closure, $\mathscr{A}$, which is called the *quasilocal algebra*. The $\mathscr{A}(\mathcal{O})$ algebra is also supposed to satisfy the analogue of local commutativity

III. $\mathcal{O}_1$ space-like separated from $\mathcal{O}_2$ implies $\mathscr{A}(\mathcal{O}_1) \subseteq \mathscr{A}(\mathcal{O}_2)'$ where the prime indicates the commutant in $\mathscr{A}$.

One can now define, in terms of these algebraic objects, the second fundamental notion of this theory, the notion of *state*. A state, ρ, is a complex-valued linear function on the quasilocal algebra $\mathscr{A}$:

$$\rho(\lambda A) = \lambda\rho(A) \qquad \text{for all} \quad \lambda \in \mathbf{C} \quad \text{and} \quad A \in \mathscr{A},$$

$$\rho(A+B) = \rho(A) + \rho(B),$$

which is positive in the sense that

$$\rho(A^*A) \geqslant 0$$

and normalized in the sense that

$$\rho(\mathbb{1}) = 1$$

where $\mathbb{1}$ is the identity in $\mathscr{A}$. (We are tacitly assuming all our C^* algebras contain an identity.) A state is said to be *invariant* under the Poincaré group if

$$\rho(A) = \rho(\alpha(a, \Lambda)(A))$$

for all Poincaré transformations (a, Λ) and all $A \in \mathscr{A}$. The importance of the notion of state is that a state determines a representation of $\mathscr{A}$ via the so-called GNS construction [68].

Theorem

If $\mathscr{A}$ is a C^* algebra with unit and ρ a state of $\mathscr{A}$, then there exists a Hilbert space $\mathscr{H}_\rho$, a vector $\Psi_\rho \in \mathscr{H}_\rho$, and a representation, π_ρ, of $\mathscr{A}$ by bounded operators in $\mathscr{H}_\rho$ such that

(1) $\rho(A) = (\Psi_\rho, \pi_\rho(A)\Psi_\rho) \qquad$ for all $\; A \in \mathscr{A}$,

(2) Ψ_ρ is a cyclic vector for $\pi_\rho(\mathscr{A})$.

Furthermore, if ρ is invariant under a group G in the sense that there is a representation $g \to \alpha_g$ of G by automorphisms of $\mathscr{A}$ such that $\rho(A) = \rho(\alpha_g(A))$, then there exists a representation $U_\rho(g)$ of G by unitary operators in $\mathscr{H}_\rho$ such that

(3) $U_\rho(g)\Psi_\rho = \Psi_\rho$,

(4) $\pi_\rho(\alpha_g(A)) = U_\rho(g)\pi_\rho(A)U_\rho(g)^{-1}$.

Thus, the GNS construction shows that to pass from the purely algebraic notions associated with the quasilocal algebra to a concrete representation, one need only choose a state. If the state is invariant, it will determine a representation in which there is an invariant vector representing the vacuum. If the state is not invariant, it may still be *covariant*, i.e., there may exist in $\mathscr{H}_\rho$ a unitary representation of the Poincaré group $\{a, \Lambda\} \rightarrow U(a, \Lambda)$ such that (4) holds for all g in the Poincaré group. It turns out that not all covariant states can be regarded as physically realizable. For example, the unitary representation of the Poincaré group $U_\rho(a, \Lambda)$ may not satisfy the spectral condition. However, even after the covariant states have been winnowed to leave only those that may reasonably be regarded as physically realizable, it may happen that there are inequivalent representations, representations that differ in their physical predictions. For example, they may yield a different mass and spin spectrum. What is the significance of such physically inequivalent representations? The answer proposed in [65] is that the theory in question has *superselection rules*, and the classes of physically admissible inequivalent representations label the coherent subspaces (see Chapter 1, pp. 5–6, for a discussion of superselection rules).

The general theory of [65] was further developed in a more specific context in [69–71] but the theory as it stood did not determine how the phenomena associated with unitary inequivalent representations of the algebras of observables would appear in concrete Lagrangian field theories.

The first example analyzed in detail was the theory of a massless scalar field in two-dimensional space-time [72]. The main motivation for this work was to show, using the ideas of the theory of local algebras, that a precise mathematical meaning can be given to proposals of Skyrme on the construction of fermion fields from boson fields [73].

Skyrme's idea was that in the theory of a neutral scalar boson field satisfying the sine-Gordon equation

$$\Box\varphi + (\alpha_0/\beta) :\sin\beta\varphi(x): = 0$$

there are particle-like excitations which should be fermions. Viewed in the light of what has come after, his exploratory calculations seem extraordinarily prescient.

What was done in [72] was to consider the very special case in which $\alpha_0 = 0$. The first step was to construct a Fock space representation of the scalar field φ satisfying

$$\Box\varphi(x) = 0.$$

To avoid troubles with infrared divergences the test functions for the field were restricted to those elements of $\mathscr{S}$ whose Fourier transforms vanish at the origin. Associated with this field φ are two conserved currents

$$\sqrt{\pi}\, j^\mu(x) = -\epsilon^{\mu\nu}\partial_\nu\varphi(x) \quad\text{and}\quad j^{5\mu}(x) = \epsilon^{\mu\nu}j_\nu(x)$$

whose corresponding charges

$$Q = \int dx\, j^0(x) \quad\text{and}\quad Q^5 = \int dx\, j^{50}(x)$$

vanish in the Fock representation. However, the local algebras constructed from φ have a two-parameter family of representations for which these charges take all real values. If one selects the subset of the representations with integer charge $\{\Pi_{n_1,n_2}$ in the Hilbert space $\mathscr{H}_{n_1,n_2}$; n_1, n_2 integers$\}$ and forms the direct sum $\bigoplus_{n_1,n_2}^{\infty}\Pi_{n_1,n_2}$ in the direct-sum Hilbert space $\mathscr{H} = \bigoplus_{(n_1,n_2)-\infty}^{\infty}\mathscr{H}_{n_1,n_2}$, then one can define charged fermion operators in $\mathscr{H}$, which decrease one or the other of the charges by one unit. At first, it was not recognized that these fermion operators may be chosen to be the two components of a fermion field satisfying the equations of the massless Thirring model, even though the natural occurrence of a free massless boson field† in the Thirring model had been known for some time [74]. However, by 1973, the construction had been explicitly carried out and recognized as a rigorous version of Skyrme's idea for the case in which $\alpha_0 = 0$ [74, 75].

For $\alpha_0 \neq 0$, it was [77] that conveyed the decisive insight. It was pointed out there that the sine-Gordon theory with general α_0 is a subtheory of a theory of fermions described by a charged two-component spinor field which satisfies the equations of the so-called massive Thirring model

$$(-i\gamma^\mu\partial_\mu + m)\psi = g\,{:}\psi^+\gamma^\mu\psi{:}\,\gamma_\mu\psi.$$

ψ is related to φ by

$${:}\psi^+(x)\gamma^\mu\psi(x){:} = -\frac{\beta}{2\pi}\epsilon^{\mu\nu}\partial_\nu\varphi, \qquad {:}\psi^+(x)\psi(x){:} = \frac{-\alpha_0}{\beta^2 m}\,{:}\cos\beta\varphi{:}$$

where $4\pi/\beta^2 = 1 + g/\pi$.

† The currents of the Thirring model are obtained from the free currents, above, by an automorphism mixing j and j^5.

In the massive Thirring model the charge Q defined (with a slightly different normalization) as

$$Q = \int dx :\psi^+(x)\gamma^0\psi(x):$$

remains a conserved quantity, but the Q^5 of the massless theory has no conserved analogue. The Hilbert space of the massive Thirring model is a direct sum

$$\mathscr{H} = \bigoplus_q \mathscr{H}_q$$

of subspaces $\mathscr{H}_q$ where q are the eigenvalues of Q, integer multiples of the charge of a single fermion. The Hilbert space of the sine-Gordon theory is the vacuum sector $\mathscr{H}_0$ of $\mathscr{H}$.

The rigorous construction of the new sectors for general α_0 and a range of β: $0 \leqslant \beta^2 \leqslant 2\sqrt{\pi}$ is found in [78–82]. The basic idea is similar to that of [72]: one constructs an automorphism of the local algebra of observables $\mathscr{A}$ which, roughly described, has the property

$$\lim_{x\to-\infty} (\Psi, \alpha_g(\varphi(x))\Psi) = \lim_{x\to-\infty} (\Psi, \varphi(x)\Psi),$$

$$\lim_{x\to+\infty} (\Psi, \alpha_g(\varphi(x))\Psi) = \lim_{x\to+\infty} [(\Psi, \varphi(x)\Psi)+g(x)]$$

for normalized states Ψ of the sine-Gordon theory itself. Then a new state ω_g on the algebra of observables $\mathscr{A}$ is defined by

$$\omega_g(A) = (\Psi_0, \alpha_g(A)\Psi_0)$$

where Ψ_0 is the vacuum of the sine-Gordon theory. If g is chosen as an appropriate "kink" function approaching $2m\pi/\beta$ as $x \to +\infty$ where m is an integer, then ω_g defines a new (m-soliton) sector. Unlike the constructions for $\alpha_0 = 0$, which are elementary and explicit, those for $\alpha_0 \neq 0$ are a tour de force of constructive field theory, since one has first to prove the existence of the solution of the cutoff-free sine-Gordon theory.

The general theory of [78–81] shows that superselection sectors associated with a "topological charge" (such as the $\beta \int dx\, \partial/\partial x^1\, g = 2\pi m$ just mentioned) occur in a wide class of theories in two-dimensional space-time, including pseudo-scalar Y_2 and $(\varphi^4)_2$—in fact, whenever the dynamics of the system results in symmetry breaking with an associated discrete family of superselection sectors. These new sectors do not contain new vacuum states; rather they contain states of new particles stabilized by the topological charge they carry, together with composite states formed of the new particles and any of the old occurring in the vacuum sectors. As these papers emphasize, there are still a number of non-trivial mathematical problems to be solved before the scattering theory in the new sectors can be regarded as firmly established. What is clear is that the reasonable interpretation of the meaning of axiom IV (asymptotic completeness) has changed as a result of these developments. If, for example, one considers the $\lambda(\varphi^4)_2$ model, one would expect asymptotic completeness to hold in the one-phase region of λ with scattering states formed from the single-particle states established by the now standard results of constructive field theory. (To be sure that there are not additional bound states one needs deep results on the spectrum of the theory [83–85].) However, for λ in the two-phase region, one expects asymptotic completeness to hold only after one has adjoined to the Hilbert space the new sectors containing a soliton or anti-soliton.

It appears that the sine-Gordon model is not a good example on which to base arguments for the necessity of this altered interpretation of asymptotic completeness, because no inelastic processes take place [86]. If that is indeed the case, meson collisions do not produce soliton–anti-soliton pairs and the scattering matrix for mesons is unitary in the vacuum sector alone. Asymptotic completeness would be expected to hold for the theory of the boson field in the vacuum sector. On the other hand, for the $(\varphi^4)_2$ theory in the two-phase region one expects colliding mesons to produce kink–anti-kink pairs, so the introduction of new sectors would be necessary to achieve asymptotic completeness.

The existence of new sectors of the $(\varphi^4)_2$ theory in the two-phase region raises some problems of interpretation: What should the Hilbert space of the theory be? Is the resulting theory a quantum field theory in the sense of the definitions of this book?

To answer the first question it is necessary to examine the physical properties of the various sectors. The two vacuum sectors $\mathscr{H}_{0+}$ and $\mathscr{H}_{0-}$ are distinguished by $\langle\varphi(x)\rangle_{0+} > 0$ and $\langle\varphi(x)\rangle_{0-} < 0$ with $\langle\varphi(x)\rangle_{0+} = -\langle\varphi(x)\rangle_{0-}$ constant in x. There is also a soliton (kink) sector, $\mathscr{H}_s$, characterized by

$$\lim_{x\to\pm\infty} \langle\varphi(x)\rangle_s = \langle\varphi(x)\rangle_{0\pm}$$

and an anti-soliton sector, $\mathscr{H}_{\bar{s}}$, characterized by

$$\lim_{x\to\pm\infty} \langle\varphi(x)\rangle_{\bar{s}} = \langle\varphi(x)\rangle_{0\mp}.$$

Notice that the soliton and anti-soliton states are not translation invariant. However, it turns out that they are covariant. The iteration of the automorphism associated with the soliton, s, on the vacuum sector $\mathscr{H}_{0,\pm}$ yields a state which again lies in $\mathscr{H}_{0,\pm}$ (symbolically $s^2 \sim e$, in the group of automorphisms) and similarly for $\bar{s}$: $\bar{s}^2 \sim e$. On the other hand, the composition, $s \circ \bar{s}$ of the automorphisms s and $\bar{s}$ applied on $\mathscr{H}_{0,\pm}$ yields a state in $\mathscr{H}_{0,\mp}$. An intuitive picture of these facts can be obtained from Figure A.3.

The standard approach to the interpretation of the four sectors $\mathscr{H}_{0+}$, $\mathscr{H}_{0-}$, $\mathscr{H}_s$, and $\mathscr{H}_{\bar{s}}$ would regard the two vacuum sectors $\mathscr{H}_{0+}$ and $\mathscr{H}_{0-}$ as physical alternatives and would attempt to adjoin $\mathscr{H}_s$ or $\mathscr{H}_{\bar{s}}$, or both, to obtain an asymptotically complete theory. If, as has been argued in [87], the lowest part of the mass spectrum in $\mathscr{H}_s$ is an isolated eigenvalue, one can use the Haag–Ruelle theory of collisions to prove the existence of kink–anti-kink states in $\mathscr{H}_{0+}$ (and in $\mathscr{H}_{0-}$) which are not meson states. Thus, to obtain an asymptotically complete theory starting from $\mathscr{H}_{0+}$ one has at least to adjoin $\mathscr{H}_s$ or $\mathscr{H}_{\bar{s}}$ or both, thereby obtaining

$$\mathscr{H}_{0+} \oplus \mathscr{H}_s \quad \text{or} \quad \mathscr{H}_{0+} \oplus \mathscr{H}_{\bar{s}} \quad \text{or} \quad \mathscr{H}_{0+} \oplus \mathscr{H}_s \oplus \mathscr{H}_{\bar{s}}$$

as state space. Similarly, starting from $\mathscr{H}_{0-}$, one has the three possibilities

$$\mathscr{H}_{0-} \oplus \mathscr{H}_s \quad \text{or} \quad \mathscr{H}_{0-} \oplus \mathscr{H}_{\bar{s}} \quad \text{or} \quad \mathscr{H}_{0-} \oplus \mathscr{H}_s \oplus \mathscr{H}_{\bar{s}}.$$

The third possibility in each case is distinguished by the existence of a parity operator; the first two are analogous to theories of neutrinos without anti-neutrinos in having a particle s or $\bar{s}$ whose anti-particle is, in fact, absent but would appear if a parity operator, P, existed and acted on the particle: $Ps = \bar{s}$.

To make further distinctions among the possibilities it is natural to associate fields with the particles s (and $\bar{s}$). The vacuum, Ψ_{0+}, is not a cyclic vector for any of the three possible Hilbert spaces which contain $\mathscr{H}_{0+}$, so that to obtain a field theory it is necessary to adjoin fields which map the vacuum sector into the other sectors. No local fields are known which do this, but by following the ideas of the theory of local algebras, one can obtain fields $s(x)$ and $\bar{s}(x)$ which have some weakened locality properties [79]. Namely, $s(x)$ anti-commutes with $\varphi(y)$ if y is sufficiently far to the right of x and commutes if it is sufficiently far to the left. However, one cannot realize both $s(x)$ and $\bar{s}(x)$ in any of the six Hilbert spaces above. Instead one needs

$$\mathscr{H}_{0+} \oplus \mathscr{H}_{0-} \oplus \mathscr{H}_s \oplus \mathscr{H}_{\bar{s}}.$$

The reason for this is that $s(x)\bar{s}(y)\Psi_{0+}$ lies in the subspace $\mathscr{H}_{0-}$, as it must if the action of $s(x)\bar{s}(y)$ implements the automorphism $s \circ \bar{s}$. (See Figure A.3.)

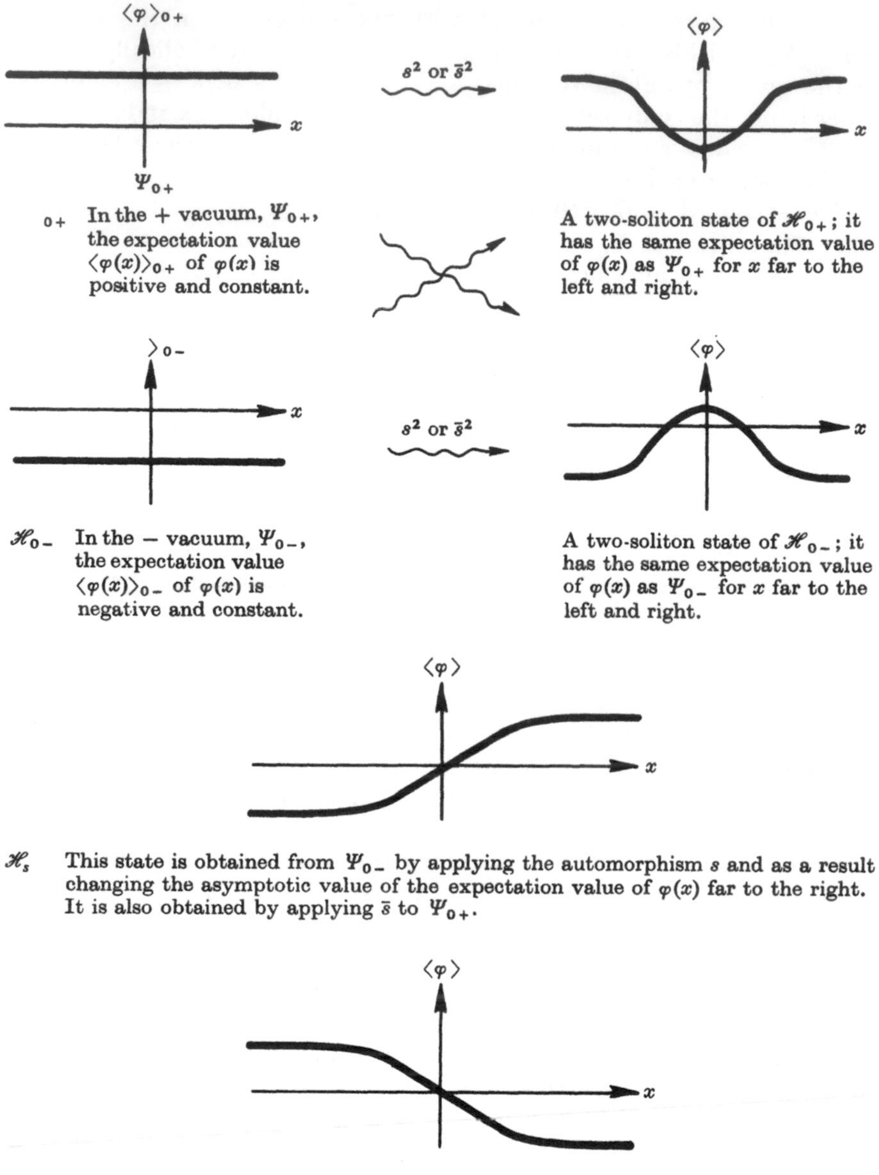

0+ In the + vacuum, Ψ_{0+}, the expectation value $\langle\varphi(x)\rangle_{0+}$ of $\varphi(x)$ is positive and constant.

A two-soliton state of $\mathscr{H}_{0+}$; it has the same expectation value of $\varphi(x)$ as Ψ_{0+} for x far to the left and right.

$\mathscr{H}_{0-}$ In the − vacuum, Ψ_{0-}, the expectation value $\langle\varphi(x)\rangle_{0-}$ of $\varphi(x)$ is negative and constant.

A two-soliton state of $\mathscr{H}_{0-}$; it has the same expectation value of $\varphi(x)$ as Ψ_{0-} for x far to the left and right.

$\mathscr{H}_s$ This state is obtained from Ψ_{0-} by applying the automorphism s and as a result changing the asymptotic value of the expectation value of $\varphi(x)$ far to the right. It is also obtained by applying $\bar{s}$ to Ψ_{0+}.

$\mathscr{H}_{\bar{s}}$ This state is obtained from Ψ_{0+} by applying the automorphism, s, and therefore changing the asymptotic value of the expectation value of $\varphi(x)$ far to the right. It is also obtained by applying $\bar{s}$ to Ψ_{0-}.

Figure A.3. Behavior of $\langle\varphi(x)\rangle$ as a function of space coordinate for some typical states of $\lambda(\varphi^4)_2$ in the two-phase region. The action of the automorphisms $s^2 = s \circ s$, $\bar{s}^2 = \bar{s} \circ \bar{s}$ and $s \circ \bar{s}$ are indicated by the wavy arrows. In particular, note that the action of the composite automorphism $s \circ \bar{s}$ is to change the asymptotic value of the expectation value of $\varphi(x)$ *both* to the right and left.

Thus, if one proposes to use $\varphi(x)$ and $s(x)$ as basic fields, one can use the Hilbert space $\mathscr{H}_{0+} + \mathscr{H}_s$, and if one uses $\varphi(x)$ and $\bar{s}(x)$, one can use $\mathscr{H}_{0-} \oplus \mathscr{H}_{\bar{s}}$; if one wants $\varphi(x)$, $s(x)$, $\bar{s}(x)$ as basic fields, one has no choice but to use $\mathscr{H}_{0+} \oplus \mathscr{H}_{0-} \oplus \mathscr{H}_s \oplus \mathscr{H}_{\bar{s}}$. The physical distinction involved is not as great as it might seem at first sight. On the one hand, the double degeneracy of the vacuum in the latter case is in principle observable. On the other hand, since the energy per unit volume of the vacuum states $\Psi_{0\pm}$ is the same, one can construct states of energy only a bit greater than $(M_s + M_{\bar{s}})c^2$ which lie in $\mathscr{H}_{0+}$ but look like Ψ_{0-} in an arbitrarily large region.

Our enumeration of the physical possibilities would not be complete without mention of the theory in which the Hilbert space is $\mathscr{H}_{0+} \oplus \mathscr{H}_s$ and the fields are $:\varphi^2:(x)$ and $s(x)$. This theory is isomorphic to the one with Hilbert space $\mathscr{H}_{0-} \oplus \mathscr{H}_{\bar{s}}$ and fields $:\varphi^2:(x)$ and $\bar{s}(x)$. In effect, by excluding observables odd in $\varphi(x)$, one has made $\varphi \to -\varphi$ into a gauge transformation. It is reasonable to conjecture that it is this theory whose observables are expressible in terms of the S-matrix.

Whatever the choice that is made among these possibilities, it appears that one may have to accept a slightly generalized notion of relativistic quantum field theory. That would not be contrary to the spirit of the present book. We already know that to describe the quantum electrodynamics of particles of spin $\frac{1}{2}$, we need field theories satisfying generalizations of axioms 0, I, II, III, and IV. The same is true for the non-Abelian gauge theories proposed to describe the weak and strong interactions. Axioms are principally useful in providing efficient guides to clear thinking, and should be changed for good and sufficient reasons.

Since the appearance of the second (1977) edition of our book, several systematic accounts of the approach to quantum field theory using local algebras have been published [88] [89], as well as reviews of further developments in the general theory of quantized fields [90] [91].

BIBLIOGRAPHY

1. R. F. Streater, "Outline of Axiomatic Relativistic Quantum Field Theory," *Rep. Prog. Phys.*, **38**, 771–846 (1975).
2. N. Bogolubov, A. Logunov, and I. Todorov, *Introduction to Axiomatic Quantum Field Theory*, W. A. Benjamin, Advanced Book Program, Reading, Mass., 1975.
3. K. Friedrichs, *Mathematical Aspects of the Quantum Theory of Fields*, Interscience, New York, 1953.

4. I. E. Segal, *Mathematical Problems of Relativistic Physics* (Lectures in Applied Math., Vol. II), Amer. Math. Soc., Providence, R.I., 1963.
5. K. Friedrichs, *Perturbation of Spectra in Hilbert Space* (Lectures in Applied Math., Vol. III), Amer. Math. Soc., Providence, R.I., 1965.
6. A. Wightman, "Introduction to Some Aspects of the Relativistic Dynamics of Quantized Fields," pp. 171–291 in *High Energy Electromagnetic Interactions and Field Theory* (M. Levy, ed.), Cargèse Lectures in Theoretical Physics, 1964, Gordon and Breach, N.Y., 1967.
7. A. Wightman, "Hilbert's Sixth Problem: Mathematical Treatment of the Axioms of Physics," pp. 147–240 in *Mathematical Developments Arising from Hilbert Problems* (Symp. in Pure Math. XXVIII), Amer. Math. Soc., Providence, R.I., 1976.
8. Y. Kato, "Some Converging Examples of Perturbation Series in Quantum Field Theory," *Prog. Theoret. Phys.*, **26**, 99–122 (1961).
9. O. Lanford, *Construction of Quantum Fields Interacting by a Cutoff Yukawa Coupling*, Ph.D. thesis (unpublished), Princeton University, Princeton, N.J., 1966.
10. T. Kato, *Perturbation Theory for Linear Operators*, Springer, New York, 1966.
11. A. Jaffe, *Dynamics of a Cutoff $\lambda\varphi^4$ Field Theory*, Ph.D. thesis (unpublished), Princeton University, Princeton, N. J., 1965.
12. A. Jaffe, O. Lanford, and A. Wightman, "A General Class of Cut-Off Model Field Theories," *Commun. Math. Phys.*, **15**, 47–68 (1969).
13. E. Nelson, "Interaction of Nonrelativistic Particles with a Quantized Scalar Field," *J. Math. Phys.*, **5**, 1190–1197 (1964).
14. J. Cannon, "Quantum Field Theoretic Properties of a Model of Nelson: Domain and Eigenvector Stability for Perturbed Linear Operators," *J. Functional Anal.*, **8**, 101–152 (1971).
15. E. Nelson, "A Quartic Interaction in Two Dimensions" in *Analysis in Function Space* (R. Goodman and I. Segal, eds.), M.I.T. Press, Cambridge, Mass., 1966.
16. A. Jaffe, "Wick Polynomials at a Fixed Time," *J. Math. Phys.*, **7**, 1250–1255 (1966).
17. J. Glimm, "Yukawa Coupling of Quantum Fields in Two Dimensions, I," *Commun. Math. Phys.*, **5**, 343–386 (1967); II, **6**, 61–76 (1967).
18. I. Segal, "Notes Toward the Construction of Non-Linear Relativisitic Quantum Fields, I: The Hamiltonian in Two Space-Time Dimensions as the Generator of a C^*-Automorphism Group," *Proc. Nat. Acad. Sci. U.S.A.*, **57**, 1178–1183 (1967); III: "Properties of the C^* Dynamics for a Certain Class of Interactions," *Bull. Amer. Math. Soc.*, **75**, 1390–139 (1969).
19. J. Glimm, "Boson Fields with Non-Linear Self-Interaction in Two Dimensions," *Commun. Math. Phys.*, **8**, 12–25 (1968).
20. J. Glimm, "Boson Fields with $:\varphi^4:$ Interaction in Three Dimensions," *Commun. Math. Phys.*, **10**, 1–47 (1968).
21. K. Hepp, "Renormalization Theory," pp. 429–500 in *Statistical Mechanics*

and Quantum Field Theory (*Les Houches 1970*) (C. DeWitt and R. Stora, eds.), Gordon and Breach, N.Y., 1971.

22. J. Glimm and A. Jaffe, "Positivity of the $\varphi_3{}^4$ Hamiltonian," *Fortschr. Physik* **21**, 327–376 (1973).
23. M. Guenin, "On the Interaction Picture," *Math. Phys.*, **3**, 120–132 (1966).
24. J. Glimm and A. Jaffe, "A $\lambda\varphi^4$ Quantum Field Theory Without Cutoffs," *Phys. Rev.*, **176**, 1945–1951 (1968).
25. J. Glimm and A. Jaffe, "Quantum Field Theory Models," pp. 1–108 in *Statistical Mechanics and Quantum Field Theory* (*Les Houches 1970*) (C. DeWitt and R. Stora, eds.), Gordon and Breach, N.Y., 1971.
26. J. Glimm and A. Jaffe, "Boson Quantum Field Theory Models," pp. 77–143 in *Mathematics of Contemporary Physics* (R. F. Streater, ed.), Academic Press, London, 1972.
27. L. Rosen, "A $\lambda\varphi^{2n}$ Theory without Cutoffs," *Commun. Math. Phys.*, **16**, 157–183 (1970).
28. J. Schwinger, "On the Euclidean Structure of Relativisitic Field Theory," *Proc. Nat. Acad. Sci. U.S.A.*, **44**, 956–965 (1958).
29. T. Nakano, "Quantum Field Theory in Terms of Euclidean Parameters," *Prog. Theoret. Phys.*, **21**, 241–259 (1959).
30. K. Symanzik, "Euclidean Quantum Field Theory, I: Equations for a Scalar Model," *J. Math. Phys.*, **7**, 510–525 (1966).
31. K. Symanzik, "Applications of Functional Integrals to Euclidean Quantum Field Theory," pp. 197–206 in *Analysis in Function Space* (W. T. Martin and I. E. Segal, eds.), M.I.T. Press, Cambridge, Mass., 1964.
32. K. Symanzik, "Euclidean Quantum Field Theory," pp. 153–226 in *Local Quantum Theory* (R. Jost, ed.), Academic Press, New York, 1969.
33. M. Reed, "Functional Analysis and Probability Theory," pp. 2–43 in *Constructive Quantum Field Theory* (Lecture Notes in Physics #25) (G. Velo and A. S. Wightman, eds.), Springer, New York, 1973.
34. E. Nelson, "Construction of Quantum Fields from Markoff Fields," *J. Functional Anal.*, **12**, 97–112; "The Free Markoff Field," 211–227 (1973).
35. F. Guerra, "Uniqueness of the Vacuum Energy Density and van Hove Phenomenon in the Infinite Volume Limit for Two Dimensional Self-Coupled Bose Fields," *Phys. Rev. Letters*, **28**, 1213–1214 (1972).
36. K. Osterwalder and R. Schrader, "Axioms for Euclidean Green's Functions," *Commun. Math. Phys.*, **31**, 83–112; **42**, 281–305 (1975).
37. E. Nelson, "Probability Theory and Euclidean Field Theory," pp. 94–124 in *Constructive Quantum Field Theory* (Lecture Notes in Physics #25) (G. Velo and A. S. Wightman, eds.), Springer, New York, 1973.
38. F. Guerra, L. Rosen, and B. Simon, "The $P(\varphi)_2$ Euclidean Quantum Field Theory as Classical Statistical Mechanics," *Ann. of Math.*, **101**, 111–259 (1975).
39. T. Spencer, "The Mass Gap for the $P(\varphi)_2$ Quantum Field Theory Model with a Strong External Field," *Commun. Math. Phys.*, **39**, 63–76 (1974)

40. B. Simon, "Correlation Inequalities and the Mass Gap in $P(\varphi)_2$, II: Uniqueness of the Vacuum for a Class of Strongly Coupled Theories," *Ann. of Math.*, **101**, 260–267 (1975).
41. B. Simon and R. B. Griffiths, "The $(\varphi^4)_2$ Field Theory as a Classical Ising Model," *Commun. Math. Phys.*, **33**, 145–164 (1973).
42. J. Glimm, A. Jaffe, and T. Spencer, "Phase Transitions for ${\varphi_2}^4$ Quantum Fields," *Commun. Math. Phys.*, **45**, 203–216 (1975).
43. G. Velo and A. S. Wightman, eds., *Constructive Quantum Field Theory* (Lecture Notes in Physics #25), Springer, New York, 1973.
44. B. Simon, *The Euclidean $P(\varphi)_2$ (Quantum) Field Theory*, Princeton University Press, Princeton, N.J., 1974.
45. J. Glimm, A. Jaffe, and T. Spencer, "The Wightman Axioms and Particle Structure in the $P(\varphi)_2$ Quantum Field Model," *Ann. of Math.*, **100**, 585–632 (1974).
46. J. Fröhlich, "Schwinger Functions and Their Generating Functionals, I," *Helv. Phys. Acta*, **47**, 265–306 (1974).
47. J. Fröhlich, "Verification of Axioms for Euclidean and Relativistic Fields and Haag's Theorem in a Class of $P(\varphi)_2$ Models," *Ann. Inst. Henri Poincaré*, **21**, 271–317 (1974).
48. R. Haag, "Quantum Field Theories with Composite Particles and Asymptotic Conditions," *Phys. Rev.*, **112**, 669–673 (1958).
49. D. Ruelle, "On the Asymptotic Condition in Quantum Field Theory," *Helv. Phys. Acta*, **35**, 147 (1962).
50. J. Dimock, "Asymptotic Perturbation Expansion in the $P(\varphi)_2$ Quantum Field Theory," *Commun. Math. Phys*, **35**, 347–356 (1974).
51. J. Eckmann, J. Magnen, and R. Sénéor, "Decay Properties and Borel Summability for the Schwinger Functions in $P(\varphi)_2$ Theories," *Commun. Math. Phys.*, **39**, 251–271 (1975).
52. J. Dimock, "The $P(\varphi)_2$ Green's Functions: Asymptotic Perturbation Expansion," *Helv. Phys. Acta*, **49**, 199–216 (1976).
53. K. Osterwalder and R. Sénéor, "The Scattering Matrix Is Non-Trivial for Weakly Coupled $P(\varphi)_2$ Models," *Helv. Phys. Acta*, **49** (1976).
54. J. P. Eckmann, H. Epstein, and J. Fröhlich, "Asymptotic Perturbation Expansion for the S-matrix and the Definition of Time-Ordered Functions in Relativistic Quantum Field Theory Models," *Ann. Inst. Henri Poincaré*, **25**, 1–34 (1976).
55. J. Glimm and A. Jaffe, "Self-Adjointness of the Yukawa$_2$ Hamiltonian," *Ann. Phys.*, **60**, 321–383 (1970).
56. R. Schrader, "Yukawa Field Theory in Two-Dimensional Space-Time without Cutoffs," *Ann. Phys.*, **70**, 412–457 (1972).
57. E. Seiler, "Schwinger Functions for the Yukawa Model in Two Dimensions with Space-Time Cutoff," *Commun. Math. Phys.*, **42**, 163–182 (1975).
58. J. Magnen and R. Sénéor, "The Wightman Axioms for the Weakly Coupled Yukawa Model in Two Dimensions," *Commun. Math. Phys.*, **51**, 297–314 (1976).

59. A. Cooper and L. Rosen, "The Weakly Coupled Yukawa_2 Field Theory: Cluster Expansion and Wightman Axioms," *Trans. Amer. Math. Soc.*, to appear.
60. E. Seiler and B. Simon, "Nelson's Symmetry and All That in the Y_2 and $(\varphi^4)_3$ Quantum Field Theories," *Ann. Phys.*, **97**, 470–518 (1976).
61. J. Feldman and K. Osterwalder, "The Wightman Axioms and the Mass Gap for Weakly Coupled $(\varphi^4)_3$ Quantum Field Theories," *Ann. Phys.*, **97**, 80–135 (1976).
62. J. Magnen and R. Sénéor, "The Infinite Volume Limit of the $\varphi_3{}^4$ Model," *Ann. Inst. Henri Poincaré*, **24**, 95–159 (1976).
63. J. Glimm and A. Jaffe, "Critical Problems in Quantum Fields," pp. 157–174 in *Les Méthodes Mathématiques de la Théorie Quantique des Champs*, CNRS, Paris, 1976.
64. R. Schrader, "A Possible Constructive Approach to $(\varphi^4)_4$, I," *Commun. Math. Phys.*, **49**, 131–153 (1976); III, **50**, 97–102 (1976); II, *Ann. Inst. Henri Poincaré*, **25** (1976).
65. R. Haag and D. Kastler, "An Algebraic Approach to Quantum Field Theory," *J. Math. Phys.*, **5**, 848–861 (1964).
66. See, however, W. Driessler and J. Fröhlich, "The Reconstruction of Local Observable Algebras from the Euclidean Green's Functions of a Relativistic Quantum Field Theory," *Ann. Inst. Henri Poincaré*, **25**, 221-136 (1977)
67. G. Emch, *Algebraic Methods in Statistical Mechanics and Quantum Field Theory*, Interscience, New York, 1972.
68. In addition to the standard accounts of the GNS construction in [47, 48] the reader may find helpful A. S. Wightman, "Constructive Quantum Field Theory: Introduction to the Problems," pp. 46–53 in *Fundamental Interactions in Physics and Astrophysics* (Iverson, Perlmutter, and Mintz, eds.), Plenum, New York, 1973.
69. H. J. Borchers, "Local Rings and the Connection of Spin with Statistics," *Commun. Math. Phys.*, **1**, 281–307 (1965).
70. S. Doplicher, R. Haag, and J. Roberts, "Field Observables and Gauge Transformations, I," *Commun. Math. Phys.*, **13**, 1–23 (1969); II, **15** 173–200 (1969).
71. S. Doplicher, R. Haag, and J. Roberts, "Local Observables and Particle Statistics," *Commun. Math. Phys.*, **23**, 199–230 (1971).
72. R. F. Streater and I. F. Wilde, "Fermion States of a Bose Field," *Nuclear, Phys.*, **B24**, 561–575 (1970).
73. T. H. R. Skyrme, "Particle State of a Quantized Meson Field," *Proc. Roy. Soc. London*, **A262**, 237–245 (1961).
74. E. Lieb and D. Mattis, "Exact Solution of a Many-Fermion System and Its Associated Boson Field," *J. Math. Phys.*, **6**, 304–312 (1965).
75. G. F. Dell'Antonio, Y. Frishman, and D. Zwanziger, "Thirring Model in Terms of Currents; Solution and Light Cone Expansion," *Phys. Rev.*, **D6**, 988–1007 (1972).
76. R. F. Streater, "Gauge Fields and Superselection Rules," *Acta Phys.*

Austriaca Suppl., **11**, 317–340 (1973); and "Charges and Currents in the Thirring Model," pp. 375–386 in *Physical Reality and Mathematical Description* (C. P. Enz and J. Mehra, eds.), D. Reidel, Dordrecht, 1974.

77. S. Coleman, "Quantum Sine-Gordon Equation as Massive Thirring Model," *Phys. Rev.*, **D11**, 2088–2097 (1975).
78. J. Fröhlich, "Poetic Phenomena in Two Dimensional Quantum Field Theory: Non-Uniqueness of Vacuum, The Solitons and All That," pp. 111–130 in *Les Méthodes Mathématiques de la Théorie Quantique des Champs*, CNRS, Paris, 1976.
79. J. Fröhlich, "New Superselection Sectors ('Soliton States') in Two Di mensional Bose Quantum Field Theories," *Commun. Math. Phys.*, **47**, 269–310 (1976).
80. J. Fröhlich, "The Pure Phases, the Irreducible Quantum Fields, and Dynamical Symmetry Breaking in Symanzik–Nelson Positive Quantum Field Theories," *Ann. Phys.*, **97**, 1–54 (1976).
81. J. Fröhlich, "Phase Transitions, Goldstone Bosons, and Topological Superselection Rules," *Acta Phys. Austriaca Suppl.*, **15**, 79–85 (1976).
82. For further general information on solitons and their significance, S. Coleman, *Classical Lumps and Their Quantum Descendants* (1975 Erice Lectures, Int. School of Subnuclear Physics "Ettore Majorana").
83. T. Spencer, "The Absence of Even Bound States for $(\varphi^4)_2$ Quantum Fields," *Commun. Math. Phys.*, **39**, 77–79 (1974).
84. T. Spencer, "The Decay of the Bethe-Salpeter Kernel in $P(\varphi)_2$," *Commun. Math. Phys.*, **44**, 143–164 (1975).
85. T. Spencer and F. Zirilli, "Scattering States and Bound States in $P(\varphi)_2$," *Commun. Math. Phys.*, **49**, 1–16 (1976).
86. M. Lüscher, *Dynamical Charges in the Quantized Renormalized Massive Thirring Model*, DESY Preprint 76/31, June 1976.
87. J. Bellissard, J. Fröhlich, and B. Gidas, "Soliton Mass and Surface Tension in the $(\lambda|\varphi|^4)_2$ Quantum Field Theory," *Phys. Rev. Letters*, **38**, 619–622 (1977).
88. R. Haag, *Local Quantum Physics*, second edition, Springer Verlag, 1996.
89. S. S. Horuzhy, *Introduction to Algebraic Quantum Field Theory*, Kluwer, 1990.
90. H. Araki, *Mathematical Theory of Quantum Fields*, Oxford, 2000.
91. N. N. Bogoliubov, A. A. Logunov, A. I. Oksak and I. T. Todorov, *General Principles of Quantum Field Theory*, Kluwer, 1990.

INDEX

Milton Keynes UK
Ingram Content Group UK Ltd.
UKHW021026140924
448309UK00006B/294

9 780691 070629